Essence of Anterior Cruciate Ligament

Konsei Shino

Essence of Anterior Cruciate Ligament

Scientific Evidence and Clinical Practice of ACL

Springer

Konsei Shino
Sports Orthopaedic Center
Yukioka Hospital
Osaka, Osaka, Japan

ISBN 978-981-99-6538-0 ISBN 978-981-99-6536-6 (eBook)
https://doi.org/10.1007/978-981-99-6536-6

This is a translated version of the book originally published in Japanese language. This has been facilitated using machine translation (by the service DeepL.com) followed by authors revising, editing and verifying the translated manuscript.

Based on a translation from the Japanese language edition: ACL no Essence Hiza Zenjujijintai no Evidence to Rinsho by Konsei Shino Copyright © Bunkodo Co., Ltd. 2022.

This Springer imprint is published by the registered company Springer Nature Singapore Pte Ltd.
The registered company address is: 152 Beach Road, #21-01/04 Gateway East, Singapore 189721, Singapore

Paper in this product is recyclable

Preface

It is no exaggeration to say that the majority of my career as an orthopedic surgeon has been spent for struggling with ACLs, since when I returned to my hometown of Osaka in my late 20s. My predecessors had not dealt with traumatic knee disorders such as ACL injuries, which were described as "internal derangements of the knee" (unknown soft tissue lesions in the knee joint that cannot be diagnosed, as the pathological diagnosis of these lesions was abandoned). Since I became interested in ACL injury, I had to find out a mentor in the literature of the United States. I had been eagerly copying the American way for 20 years until my late 40s, and finally realized their approach is a bit different from the real anatomy. Thus, restoring stability without loss of motion after treating ACL injury was a dream. Thereafter, I decided to restart everything to pursue the real anatomical way of treating ACL injury.

Based on the accumulated anatomical knowledges, the new and more precise anatomical ACL reconstruction techniques were developed to closely mimic the native ACL. Thanks to younger colleagues, the surgical techniques were proved to be biomechanically more efficient to restore the stability. As a result, the clinical outcomes improved drastically as well. Now, restoring stability without loss of motion after treating ACL injury is not a dream but a reality.

This book describes the progress on ACL studies accumulated by my younger colleagues and myself over the last 20 years or so. I believe this will be useful to readers.

Osaka, Osaka, Japan — Konsei Shino
July 2023

The original version of the book was revised: The copyright holder has been updated in the book. A correction to the book is available at https://doi.org/10.1007/978-981-99-6536-6

Contents

About the Author

Konsei Shino Born in Osaka in 1948

Director & Consultant, Sports Orthopaedic Center, Yukioka Hospital

Visiting Professor, Department of Orthopaedics, Sapporo Medical University

Education:

Graduated from the University of Tokyo, Faculty of Medicine in March 1973

Experience:

Department of Orthopaedics, The University of Tokyo

Department of Orthopaedic Surgery, Osaka University

Founder, Department of Orthopedic Sports Medicine, Osaka Rosai Hospital

Professor, Faculty of Comprehensive Rehabilitation, Osaka Prefecture University

Academic meetings presided:

The 28th Annual Meeting of the Japanese Arthroscopy Association

The 34th Japanese Society of Clinical Biomechanics

The 1st JOSKAS (Japanese Orthopaedic Society of Knee, Arthroscopy and Sports Medicine)

Honorary member:

AANA (Arthroscopy Association of North America)

ISAKOS (International Society of Arthroscopy, Knee Surgery and Orthopaedic Sports Medicine)

JOSKAS (Japanese Orthopaedic Society of Knee, Arthroscopy and Sports Medicine)

Japanese Society of Clinical Biomechanics

Awards:

John Joyce Award (ISAKOS) × 3

Albert Trillat Award (ISAKOS) × 2

Masaki Watanabe Award (JOSKAS)

Takagi & Watanabe Award (APKASS: Asia-Pacific Society of Knee, Arthroscopy and Sports Medicine)

Porto Award (ESSKA: European Society of Sports Traumatology, Knee Surgery and Arthroscopy)

Books:

Sports Knee Clinic [Sole author, Kanehara, 2008 (first edition), 2014 (second edition)] (in Japanese)

New Edition of Sports Orthopedics (Co-editor, Nankodo, 2011) (in Japanese)

ACL Injury and Its Treatment (co-editor, Springer, 2016)

Chapter 1
Author's Struggle with ACL Injury

1.1 Diagnosis

I first became interested in ACL injuries when I was a resident at the University of Tokyo Hospital in 1973. At that time, the diagnosis of ACL injury was based on an anterior drawer test in 90-degree flexion. However, the diagnosis of ACL injury was difficult because of the high incidence of false-negative in the test. At that time, Dr. Nakajima H, who was chief of knee and sports medicine in the Hospital, discovered the N-test (Jerk test in European and American literature) and suggested that it might be useful for the diagnosis of ACL injury. However, it was difficult to understand the usefulness of the test because of a lack of scientific proof. The reproducibility of the test was also poor due to patients' muscular defense.

In 1979, I joined Osaka University Hospital, where I was in charge of knee and sports medicine. A police officer patient suffering from instability after medial meniscectomy at another hospital came to our hospital. The patient demonstrated anterior subluxation of the tibial plateau by himself on the bed while extending and flexing the knee with abduction force applied. This was the first time I understood the phenomenon of anterior subluxation of the tibial plateau due to ACL injury (Fig. 1.1).

Shortly before that, Torg reported the usefulness of the Lachman test, an anterior drawer test in near extension [1]. At that time, when I first reported the usefulness of this test at the Japan Knee Society Meeting, I was totally opposed by the audience. Later on, the usefulness of this test was biomechanically proven in cadaveric knees by Fukubayashi, and the maneuver for detecting anterior instability due to ACL injury was established [2].

Due to poor diagnostic capability among surgeons, the patients were left undiagnosed and suffering from repeated knee subluxations/giving ways for a long time. Thus, when I finally made the correct diagnosis on them, most of them were associated with severely-torn/degenerative menisci.

K. Shino, *Essence of Anterior Cruciate Ligament*,
https://doi.org/10.1007/978-981-99-6536-6_1

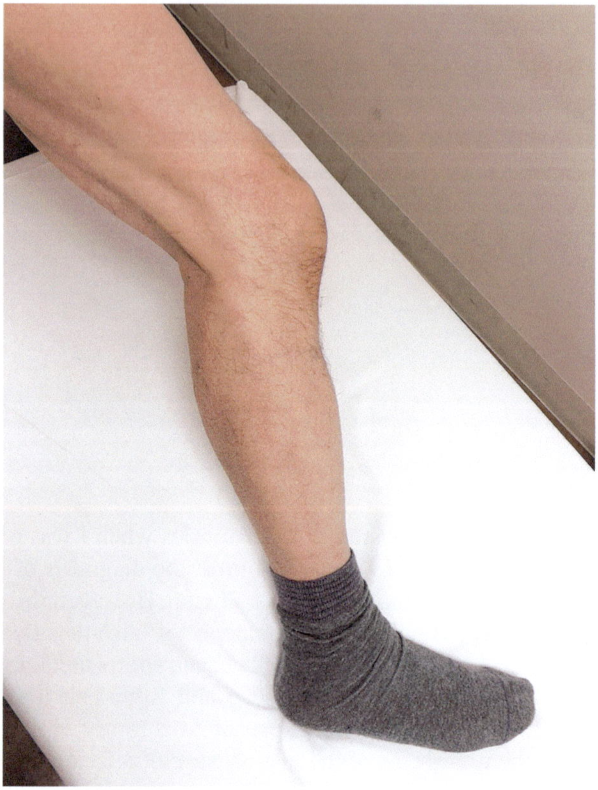

Fig. 1.1 Limb position for self-demonstrating anterior subluxation-reduction phenomenon of the tibial plateau
The patient places the medial aspect of the foot of the affected limb on the bed, and then extends the knee and flexes it with the knee-abduction force applied

However, even with the Lachman test, the influence of muscular defense was inevitable, thus the diagnosis in the acute stage was not easy. The diagnosis of ACL injury became easy around 1990 only after MRI became popular in Japan.

1.2 Surgical Treatment

Around the 1970s, the generally-recommended treatment of ACL injuries was primary repair for acute cases [3]. Therefore, when an early diagnosis was made, primary repair was performed in the acute phase. However, we were disappointed by the poor results in our own cases, although good results for athletes were reported in the Western literature. For chronic cases, as control of lateral tibial plateau subluxation/anterolateral rotatory instability was emphasized, lateral extra-articular tenodesis using the iliotibial band rather than intra-articular reconstruction was advocated and its good postoperative results in athletes were reported [4]. I immediately replicated this procedure in several cases, but the results were not good at all. Therefore, we switched to perform simultaneous intra- and extra-articular reconstruction with the iliotibial band (MacIntosh method), but this did not result in satisfactory outcomes either [5]. (Fig. 1.2).

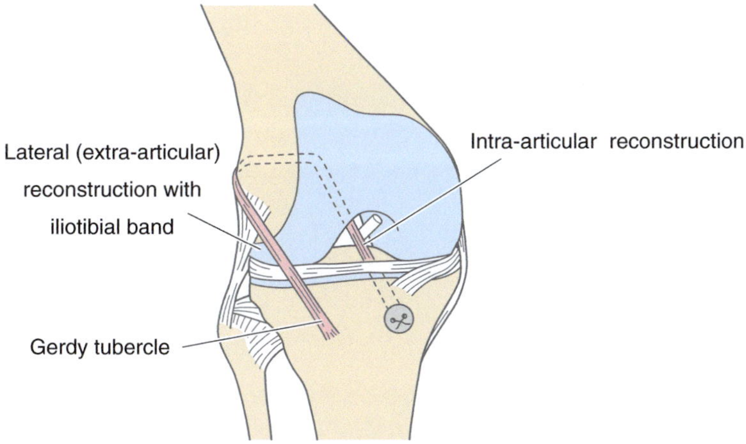

Fig. 1.2 Combined extra-articular (anterolateral) and intra-articular reconstruction using the iliotibial band (MacIntosh method/Lateral substitution over-the-top: LSOT)
The distally-attached iliotibial band, is introduced into the joint through the over-the-top (posterior superior margin of the lateral femoral condyle), and led to the tibial tunnel, and fixed with a pull-out suture over a button

Eventually the importance of intra-articular reconstruction was recognized, but the results of the generally-performed reconstructions using pedicled autologous tendon grafts were poor, while good outcomes were reported. For this reason, in the 1980s, people jumped to artificial ligaments made of polyester (Dacron), polytetrafluoroethylene (GORE-TEX), carbon fiber and so on. However, their trials resulted in failure worldwide. In 1981, the author started intra-articular reconstruction using allogeneic tendon grafts in search of non-autologous collagen tissue. Although the usefulness of allogeneic tendons as a ligament replacement material was experimentally and clinically proven, the lack of tissue banks in Japan made it impossible to continue this procedure for a number of patients. Thus the use of allogeneic tendons was discontinued.

In 1986, the surgical approach was shifted from arthrotomy to arthroscopy despite disagreement by the senior professor. Therefore, I had to purchase arthroscopic surgical instruments at my own expenses, as they were not available in the University Hospital.

Around 1987, the concept of "isometric ligament reconstruction," in which the length of the reconstructed ligament remains unchanged during knee flexion and extension movement, was introduced by Graf et al. in the United States [6].This isometric reconstruction technique, in which a femoral tunnel is created through the tibial tunnel, became popularized in Japan by the author, who had been adoring surgeons in the United States at that time. However, this technique brought the reconstructed ligament different in orientation from the normal ligament, resulting in high percentage of poor result cases [7]. (Fig. 1.3). In 1999, we began the anatomical studies to elucidate the position of the bone tunnels for anatomical reconstruction of the ligament. As a result, I devised the "bony landmark strategy" (see Chap. 7), which made it possible to accurately identify the anatomical attachment

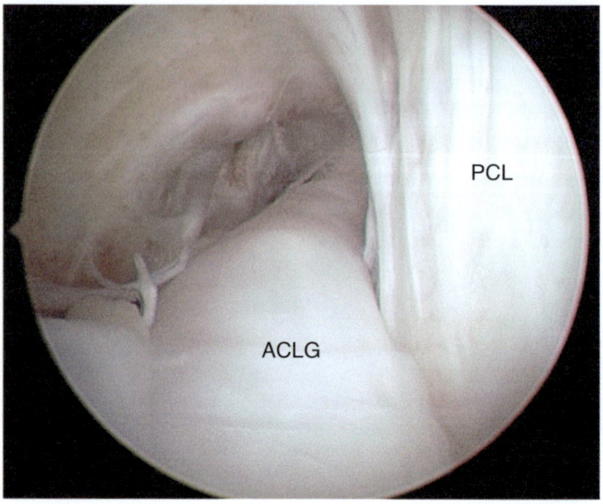

Fig. 1.3 An arthroscopic view of the reconstructed ligament with the trans-tibial tunnel technique in the right knee, based on the concept of "isometric" ligament reconstruction
The reconstructed ligament (ACLG) is pushed by the posterior cruciate ligament (PCL), and looks different in orientation from the normal ligament

site of the ligament during the arthroscopic surgery [8, 9]. Thus the "anatomical" ligament reconstruction was realized.

The reconstruction techniques using the patellar tendon or hamstring tendon to be described later, was developed to quite faithfully reproduce the morphology and orientation of the normal ligament. This has led to a dramatic improvement in results.

1.3 Rehabilitation

After an invasive surgery due to arthrotomy followed by several weeks of long leg cast immobilization, postoperative joint contracture was inevitable. At first, I was extremely embarrassed by the strong suggestion of the senior professor to apply a hip spica cast. After some "negotiation," a long leg cast was permitted to apply.

At that time, there was no distinct line between joint stiffness due to contracture and stability restored by the reconstructed ligament. It was a general consensus that limitation of range of motion was inevitable to achieve stability. Almost all of the patients were required to postoperatively undergo forcible manual mobilization under anesthesia. Restoration of a resilient and stable knee without stiffness was a dream.

The only way to prevent joint stiffness/fibrosis is to avoid long-term immobilization and to begin range of motion exercise as early as possible. Following a worldwide big wave of early postoperative range of motion exercise around 1990, we started range of motion training on the third day after the reconstruction using soft tissue grafts (without bone plugs), resulting in a drastic increase of failure cases. As the reconstructed ligament consisting of soft tissues moves within the bone tunnels during range of motion exercise, too early range of motion exercise should be avoided for the integration of the reconstructed ligament to the tunnel wall. While immobilization is advantageous for the integration of the reconstructed ligament to the tunnel wall, it also poses the risk of joint fibrosis. By carefully managing this antinomy, we were able to achieve both mobility and stability.

1.4 Lessons Learned

(a) Statements made by famous doctors are often false and exaggerated. They should not be trusted easily. The correct answer is often found in a frank observation on our patients or the anatomy.
(b) At the time of treating ligament injuries, our goal should be to achieve restoration of stability without loss of range of motion. For this purpose, ligament reconstruction or repair should be performed as closely as possible to mimic the normal ligament, and appropriate rehabilitation should be followed.

References

1. Torg JS, et al. Clinical diagnosis of anterior cruciate ligament instability in the athlete. Am J Sports Med. 1976;4:84–93.
2. Fukubayashi T, et al. An in vitro biomechanical evaluation of anterior-posterior motion of the knee. Tibial displacement, rotation, and torque. J Bone Joint Surg Am. 1982;64:258–64.
3. O'Donoghue DH. An analysis of end results of surgical treatment of major injuries to the ligaments of the knee. J Bone Joint Surg Am. 1955;37:1–13.
4. Losee RE, et al. Anterior subluxation of the lateral tibial plateau. A diagnostic test and operative repair. J Bone Joint Surg Am. 1978;60:1015–30.
5. Bertoia JT, et al. Anterior cruciate reconstruction using the MacIntosh lateral-substitution over-the-top repair. J Bone Joint Surg Am. 1985;67:1183–8.
6. Graf B. Isometric placement of substitutes for the anterior cruciate ligament. In: Jackson DW, Drez D, editors. The anterior cruciate deficient knee. St Louis: CV Mosby; 1987. p. 102–13.
7. Toritsuka Y, Shino K, et al. Second-look arthroscopy of anterior cruciate ligament grafts with multi-stranded hamstring tendons. Arthroscopy. 2004;20:287–93.
8. Iwahashi T, Shino K, et al. Direct anterior cruciate ligament insertion to the femur assessed by histology and 3-dimensional volume-rendered computed tomography. Arthroscopy. 2010;26:S13–20.
9. Otsubo H, Shino K, et al. The arrangement and the attachment areas of three ACL bundles. Knee Surg Sports Traumatol Arthrosc. 2012;20:127–34.

Chapter 2
Anatomy and Biomechanics

2.1 Characteristics of the Knee Joint

The knee joint is the largest weight-bearing joint in the human body and consists of two roller-shaped femoral condyles proximally and a nearly flat tibial plateau distally.

Since the plateau is posteriorly tilted about 11°, the gravity load acts as a force that slides the femoral condyle posteriorly (anterior tibial drawer force). In other words, the ACL is always loaded in the standing position (Fig. 2.1).

The knee joint is not conforming unlike the hip joint, support by soft tissues (ligaments and meniscus) contributes greatly to the stability of the joint. The medial tibial plateau is slightly concave, while the lateral tibial plateau is convex. The lateral meniscus has greater coverage of the tibial plateau than the medial meniscus (Fig. 2.2).

The trochlea of the femur is articulated by the patella which is the sesamoid of the quadriceps femoris femoris muscle.

K. Shino, *Essence of Anterior Cruciate Ligament*,
https://doi.org/10.1007/978-981-99-6536-6_2

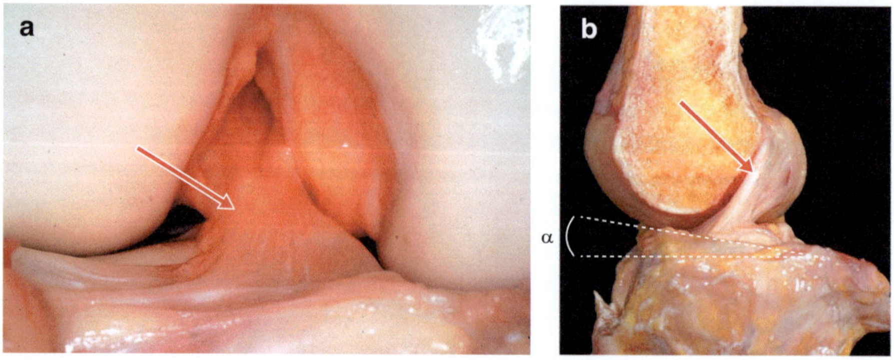

Fig. 2.1 ACL of the right knee(**a**) Frontal view (arrow), (**b**) Lateral view (arrow) (The medial femoral condyle is removed to visualize the entire length of the ACL)
The ACL attaches to the posterior superior margin of the lateral wall of the intercondylar notch of the femur, and inserts anteriorly and slightly medially to the tibial plateau. Since the tibial plateau is posteriorly inclined (α-angle), a vertical load acts as a force that slides the femoral condyle posteriorly (anterior tibial drawer force). During the heel-contact phase of a walking cycle, the increased backward tilt of the tibia results in the increase in α-angle, so the backward shear force to the femur (anterior tibial drawer force) also increases

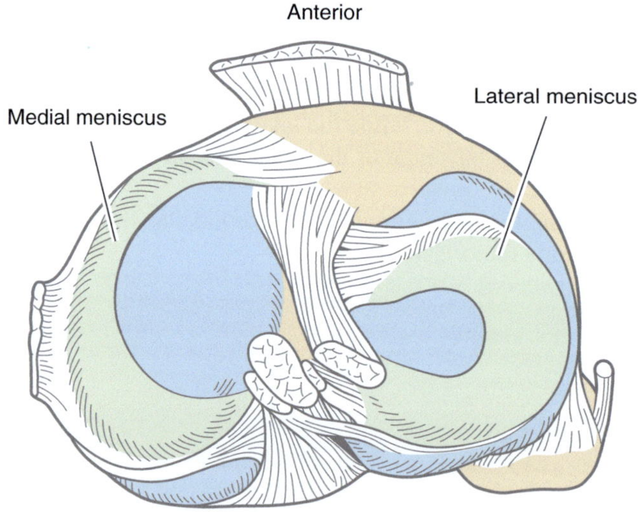

Fig. 2.2 Tibial plateau of the right knee (from above)
The lateral meniscus has greater coverage of the tibial plateau than the medial meniscus

2.2 Muscles Around the Knee Joint

Ligaments and menisci act as static stabilizer, while muscles act as dynamic stabilizer.

1. Quadriceps femoris muscle

The quadriceps femoris muscle, the largest muscle in the human body, is located anteriorly as an extensor with the patella and acts for acceleration and deceleration during sports activities.

In the sagittal plane, contraction of the quadriceps femoris muscle produces an anterior tibial drawer force in knee extension. The anterior tibial drawer force decreases with flexion, and at 60–70° (quadriceps femoris neutral angle) the anterior-posterior force becomes zero, while at 90° it becomes posterior tibial drawer force (Fig. 2.3) [1].

In the frontal plane, the contraction of the quadriceps femoris muscle in knee extension produces the tibial internal rotation torque due to the Q angle (normal value: 13~18°) (Fig. 2.4).

2. Hamstring muscles

The hamstring muscles pull the tibial plateau backward, but their force is weak in the extended position and strong when the tibia is flexed more than 60°.

3. Popliteal muscle

The popliteal muscle originates from the lateral epicondyle of the femur and posterior segment of the lateral meniscus and inserts on the posterior aspect of the proximal tibia. It functions as an internal rotator of the tibia and contributes to the posterior translation of the lateral meniscus during knee flexion.

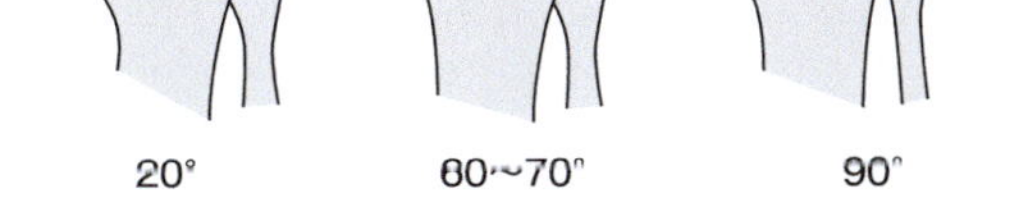

Fig. 2.3 Forces acting on the tibia due to quadriceps femoris muscle contraction (Fq)
Force by quadriceps femoris muscle contraction (Fq) to the tibia is divided into two component forces: fh and fvAt 20° of flexion/near extension, component force fh acts as the anterior tibial drawer force, and it decreases with flexion. At 60–70° (quadriceps femoris neutral angle), component force fh becomes zero. Component force fh becomes a posterior tibial drawer force at 90° (Cited from Ref. [1])

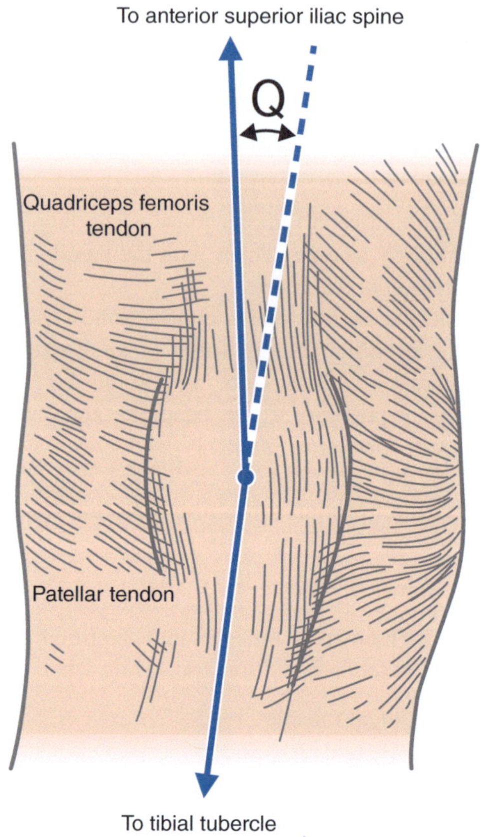

Fig. 2.4 Exertion of quadriceps femoris contraction to the tibia in knee extension (frontal plane)
Quadriceps femoris muscle contraction, due to Q angle, produces tibial internal rotation torque if the patella is stable on the femoral trochlea

2.3 Functional Anatomy of ACL

The ACL is a tape-like ligament attached to the posterior superior margin of the lateral wall of the intercondylar notch and slightly medial to the anterior tibial plateau (Fig. 2.1). It is approximately 38 mm long (longest part), 11 mm wide (longest diameter), and 3 mm thick (shortest diameter) [2, 3].

The surface is covered with synovial membrane and blood flow comes from the posterior capsule through one branch of the middle genicular artery.

1. Fiber bundles and their microstructure

The ligament had been classified into "anteromedial bundle" (AMB) and "posterolateral bundle" (PLB) based on observation in knee flexion [2].

In recent years, the conventional AMB has been further divided into medial and lateral fiber bundles, and three fiber bundle classifications have been proposed: anteromedial-medial bundle (AMMB), anteromedial-lateral bundle (AMLB), and PLB (Figs. 2.1 and 2.5) [1–13].

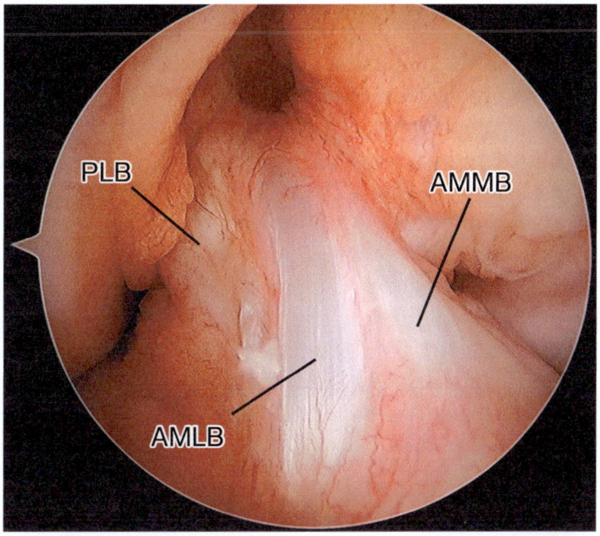

Fig. 2.5 Arthroscopic view of 3 ACL bundles in the right knee
It is divided into AMMB: anterior medial-medial bundle, AMLB: anterior medial-lateral bundle, and PLB: posterior lateral bundle

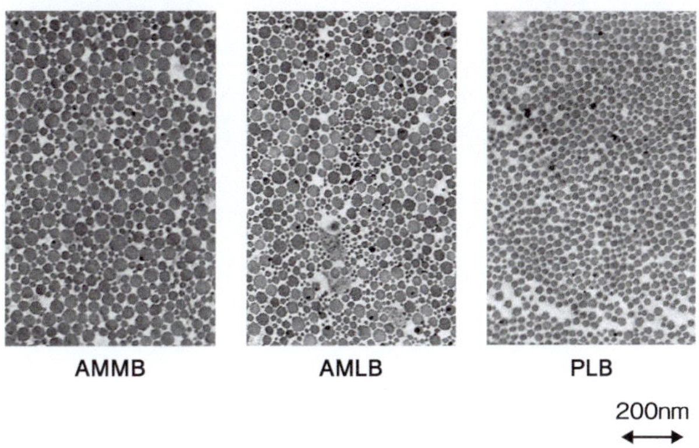

Fig. 2.6 Ultrastructure of ACL (Transmission electron microscopy findings in transverse section)
AMMB shows fibrils of various sizes (20–120 nm), while PLB is mostly composed of 50–60 nm fibrils AMLB is intermediate between the two(Adapted from Ref. [4])

The anterior two bundles: AMMB and AMLB are tense in almost entire range of motion, while the posterior PLB becomes slack in flexion.

The ultrastructure varies with each fiber bundle (Fig. 2.6) [4]. AMMB shows a gradual bimodal distribution of collagen fibrils with various sizes ranging from 20 to 120 nm in diameter. PLB shows unimodal distribution of smaller diameter fibrils of 50–60 nm [4]. AMLB shows an intermediate structure between the two. This ultrastructure is reflected in the mechanical properties described below [5].

POINT
Ligaments with strong mechanical properties show a bimodal distribution of collagen fibrils, whereas those with poor mechanical properties show a unimodal distribution, with collagen fibrils of 50–60 nm in diameter occupying most of the area.

2. Attachment areas (Figs. 2.7 and 2.8)

The location of the femoral attachment area was histologically revealed by Iwahashi et al. and projected on a 3-D CT (Fig. 2.7) [6]. The location of the tibial attachment area was elucidated by Tensho, Kusano et al. (Fig. 2.8) [7, 8].

Furthermore, the attachment areas of the 3-fiber bundle were demonstrated by Otsubo et al. [9] (Figs. 2.7 [6, 9] and 2.8 [7, 8]).

Unlike the tibial attachment of the medial collateral ligament, the ACL attachments are the direct insertion type with a fibrocartilaginous layer. Therefore, the bony surface of the attachment area is slightly concave (Fig. 2.9).

On the femoral side, the angular change of the ligament-femoral axis during flexion and extension is large and the fibrocartilage layer is thick. On the other hand, on the tibial side, the angular change is minimal and the fibrocartilage layer is thin.

The subchondral bone under the fibrocartilaginous layer is known to be oriented and is more pronounced posteriorly on the femoral side and anteriorly on the tibial side [10].

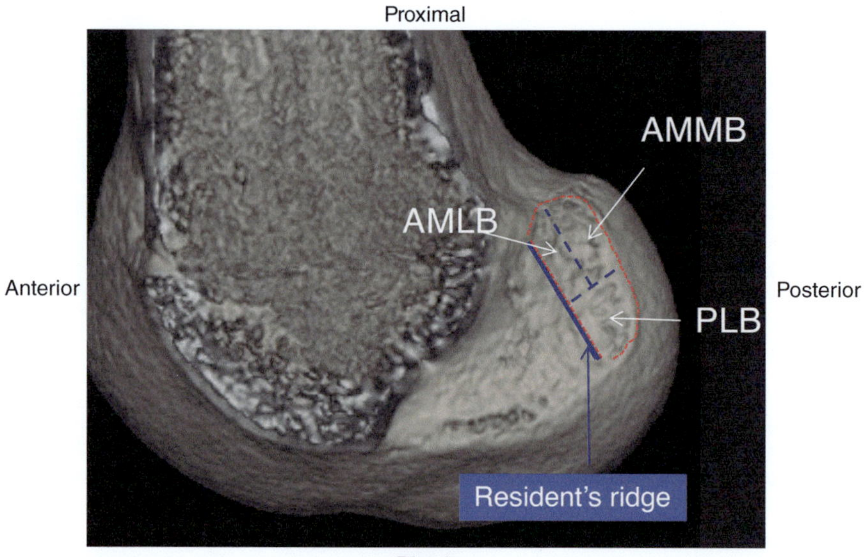

Fig. 2.7 Femoral attachment area shown on the lateral wall of the intercondylar notch on 3-D CT (area surrounded by red dashed line)
A semilunar area posterior to the bony prominence: Resident's ridge
Attachment areas of the three bundles (AMMB, AMLB, and PLB) are shown
(Adapted from Refs. [6, 9])

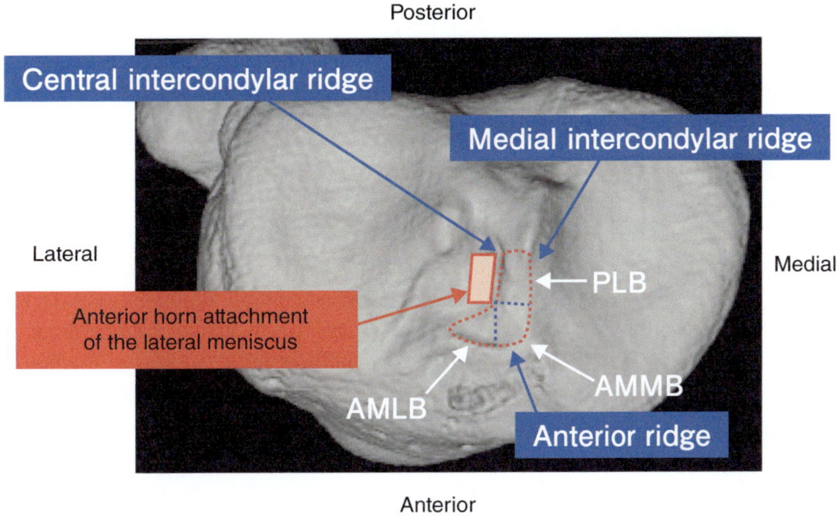

Fig. 2.8 Tibial attachment area on the right tibial plateau on 3-D CT (area surrounded by red dashed line)
It has a boot-like shape surrounded by three bony ridges: medial intercondylar ridge, anterior ridge, and central intercondylar ridge
Attachment areas of the three bundles (AMMB, AMLB, and PLB) are shown
(Adapted from Refs. [7, 8])

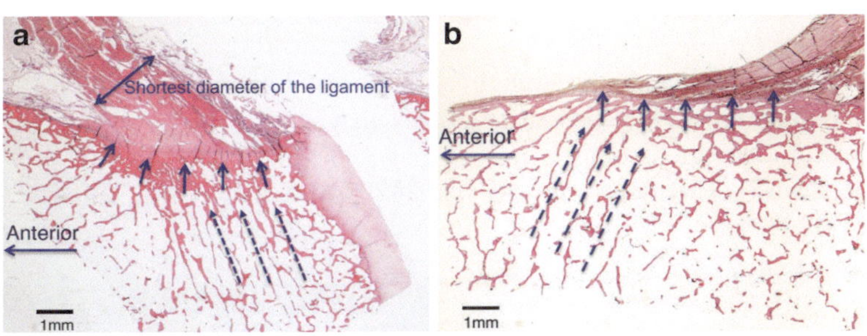

Fig. 2.9 Histology of ACL attachment areas (HE staining)
(**a**) Femur (oblique transverse section parallel to Blumensaat's line)
(**b**) Tibia (sagittal plane)
The bony surface of the attachment area is slightly concave (arrow), and the bony trabeculae under the area are oriented, more pronounced posteriorly on the femoral side (A, dashed arrow) and anteriorly on the tibial side (B, dashed arrow)
(Adapted from Ref. [10], photos courtesy of Daisuke Suzuki)

2.4 Biomechanics of the ACL

1. Biomechanical role

It plays a major role in controlling anterior subluxation of the tibial plateau as well as internal rotation of the tibia.

The functional position of the knee joint is in extension, and gravity causes the femoral condyle to slide backward on the posteriorly-inclined tibial plateau, resulting in anterior tibial drawer force during loading. The ACL is a restraint against this anterior tibial drawer force. Therefore, patients with this ligament insufficiency show an anterior/antero-lateral subluxation of the tibia during heel contact in a walking cycle. To avoid this subluxation, the patient tries to load with the knee flexed and to keep the tibial plateau anteriorly-tilted to reduce the subluxation.

2. *In situ* load to the ACL

The load on the ACL during passive flexion and extension under non-weight bearing condition was directly measured in a cadaveric knee by Markolf et al. The tension is almost zero from 45° to 10° of flexion, and gradually increases from 10° of flexion to hyperextension, reaching about 120 N (Fig. 2.10) [11].

The load applied during exercises including walking, sprinting, or jumping, reaches maximum during heel contact. The maximum load during walking is estimated to be around 300 N [12].

3. Mechanical properties

(1) Structural Properties

According to Woo et al., the strength of human ACLs decreased with age, and the average maximum tensile load was 2160 N in young subjects (average age 29), 1503 N in middle-aged subjects (average age 45), and 658 N in elderly subjects (average age 75). The mean stiffness was 242 N/mm, 220 N/mm, and 180 N/mm, respectively [13].

POINT
Age-related decrease in ACL strength leads to increased anterior tibial shift during loading due to posterior inclination of the tibial plateau, which in turn leads to increased stress on the posterior horn of the medial meniscus. This may be one of the causes of degenerative tear of the posterior horn of the medial meniscus.

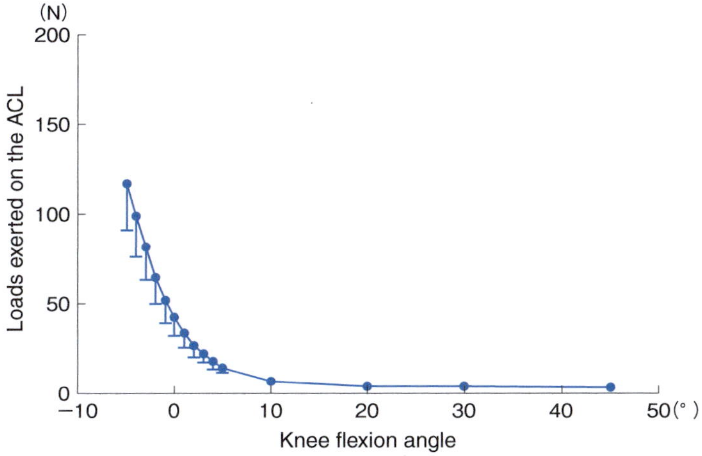

Fig. 2.10 Loads on ACL during passive flexion-extension
From 45° to 10° flexion, the ACL tension is almost 0. From 10° to hyperextension, it gradually increases to about 120 N
(Modified from Ref. [11])

(2) Material Properties

According to Butler, the average maximum tensile strength was 46 MPa for AMMB, 31 MPa for AMLB, and 15 MPa for PLB. The mean elastic modulus was 283 MPa, 286 MPa, and 155 MPa, respectively [14]. These differences in strength among the bundles correspond to the aforementioned differences in ultrastructure (Fig. 2.6).

POINT
The strongest AMMB has the longest distance between the two attachment areas. The AMMB is considered to play the greatest stabilizing function against higher loads.

(3) Role of each fiber bundle

Fujie et al. conducted a biomechanical study under 100 N anterior tibial drawer load (laboratory level load) and obtained the following results [15]. In extension,

AMMB, AMLB, and PLB are all tense to carry the load. In flexion, the PLB relaxes and the two anterior bundles, especially the AMMB, bear the main load.

POINT

When the anterior tibial load is further increased to about 500–1000 N (sports activity level load) in extension, the anterior displacement of the tibia increases and the longest bundle (AMMB), which attaches to the most posterior portion in the femoral attachment area (fan-like extension), is assumed to mainly carry the load. This assumption is supported by a biomechanical experiment in porcine model that has shown that ACL failure loads were drastically reduced after cutting the fibers attaching to the fan-like extension [16].

(4) Complementary role of the meniscus in anterior tibial drawer load

The meniscus is a C-shaped fibrocartilage that lies between the femoral condyle and tibial plateau. The inner margin is thin, while the outer periphery is thicker. The anterior part is thin and the posterior part is thicker, the outer periphery is attached to the joint capsule, and the anterior and posterior horns are attached to the tibial plateau. In particular, the posterior segment of the medial meniscus attaches to the thicker joint capsule and has a greater restraining effect against anterior tibial drawer force.

According to the experiment by Levy et al. using a low load of about 100 N, the restraining effect of the medial meniscus against anterior tibial drawer load is almost negligible when the ACL is normal [17]. This is based on the experimental data with a low anterior tibial drawer load of about 100 N without weight-bearing. The restraining effect of the posterior segment of the medial meniscus against anterior tibial drawer load with increases of the anterior tibial drawer load as well as the joint compressive load due to gravity.

POINT

In most biomechanical experiments, loads as low as 100 N are used. In actual activities, loads of several hundred Newtons are applied to the ACL even during walking. As the load increases, the positional relationship between the femur and tibia changes, and the results will differ from those of experiments using low loads.

Attention should be paid in interpreting the results of biomechanical experiments (under low loads). For instance, it should not be easily assumed that medial meniscectomy does not affect joint stability when the ACL is normal.

References

1. Daniel DM, et al. Use of quadriceps femoris active test to diagnose posterior cruciate ligament disruption and measure posterior laxity of the knee. J Bone Joint Surg Am. 1988;70:386–91.
2. Girgis FG, et al. The cruciate ligaments of the knee joint. Anatomical, functional and experimental analysis. Clin Orthop Relat Res. 1975;106:216–31.
3. Smigielski R, et al. Ribbon like appearance of the midsubstance fibres of the anterior cruciate ligament close to its femoral insertion site: a cadaveric study including 111 knees. Knee Surg Sports Traumatol Arthrosc. 2015;23:3143–50.
4. Suzuki D, Shino K, et al. Ultrastructure of the three anterior cruciate ligament bundles. Clin Anat. 2015;28:910–6.
5. Parry DA, et al. A comparison of the size distribution of collagen fibrils in connective tissues as a function of age and a possible relation between fibril size distribution and mechanical properties. Proc R Soc Lond (Biol). 1978;203:305–21.
6. Iwahashi T, Shino K, et al. Direct anterior cruciate ligament insertion to the femur assessed by histology and 3-dimensional volume-rendered. Arthroscopy. 2010;26:S13–20.
7. Tensho K, et al. Bony landmarks of the anterior cruciate ligament tibial footprint: a detailed analysis comparing 3-dimensional computed tomography. Am J Sports Med. 2014;42:1433–40.
8. Kusano M, Yonetani Y, Shino K, et al. Tibial insertions of the anterior cruciate ligament and the anterior horn of the lateral meniscus: a histological and computed tomographic study. Knee. 2017;24:782–91.
9. Otsubo H, Shino K, et al. The arrangement and the attachment areas of three ACL bundles. Knee Surg Sports Traumatol Arthrosc. 2012;20:127–34.
10. Suzuki D, Shino K, et al. Functional adaptation of the fibrocartilage and bony trabeculae at the attachment sites of the anterior cruciate ligament. Clin Anat. 2020;33:988–96.
11. Markolf KL, et al. Direct measurement of resultant forces in the anterior cruciate ligament. An in vitro study performed with a new experimental technique. J Bone Joint Surg Am. 1990;72:557–67.
12. Shelburne KB, et al. Pattern of anterior cruciate ligament force in normal walking. J Biomech. 2004;37:797–805.
13. Woo SL, et al. Tensile properties of the human femur-anterior cruciate ligament-tibia complex. The effects of specimen age and orientation. Am J Sports Med. 1991;19:217–25.
14. Buter DL, et al. Location-dependent variations in the material properties of the anterior cruciate ligament. J Biomech. 1992;25:511–8.
15. Fujie H, Shino K, et al. Mechanical functions of the three bundles consisting of the human anterior cruciate ligament. Knee Surg Sports Traumatol Arthrosc. 2011;19:S47–53.
16. Sabzevari S, et al. The femoral posterior fan-like extension of the ACL insertion increases the failure load. Knee Surg Sports Traumatol Arthrosc. 2020;28:1113–8.
17. Levy IM, et al. The effect of medial meniscectomy on anterior-posterior motion of the knee. J Bone Joint Surg Am. 1982;64:883–8.

Chapter 3
Mechanism, Epidemiology of ACL Injury

3.1 Mechanism of Injury

Injuries without contact with others often occur when landing from a jump, stopping suddenly, or changing direction. The injuries frequently occur in events that involve cutting/deceleration movements, such as basketball, apparatus gymnastics, volleyball, and handball. Falls while skiing are also frequent [1]. In school-age girls, some injuries occur during repetitive sideways jumping, which is performed as an agility test. Some cases got injured during a straight line sprint.

On the other hand, the ligament is often injured during contact sports involving physical contact, such as rugby and judo. In these cases, medial collateral ligament injuries may also be associated, but they are often mild to moderate.

This injury occurs as a result of an anterior dislocation/subluxation of the knee joint caused by the femoral condyle sliding backward over the tibial plateau. The diagnosis of dislocation is never made because the knee is immediately reduced. Specifically, it is most likely to occur when the following conditions are met: (1) The foot is caught on the ground due to the catching of mismatched (spiked) shoes, improper adjustment of the ski-binding, etc. (2) Posterior inclination of the trunk/heel-grounding increases the posterior inclination of the tibial plateau (Fig. 3.1b). (3) The quadriceps femoris muscle contracts vigorously in knee extension (Fig. 2.3). In addition, when the knee is forcibly abducted, the convex lateral femoral condyle tends to slide backward on the convex tibial plateau, leading to anterolateral dislocation of the tibia.

POINT

This injury is most likely to occur when (1) landing on the heel, (2) posterior inclination of the trunk, and (3) vigorous quadriceps contraction (plus knee abduction) in extension.

K. Shino, *Essence of Anterior Cruciate Ligament*,
https://doi.org/10.1007/978-981-99-6536-6_3

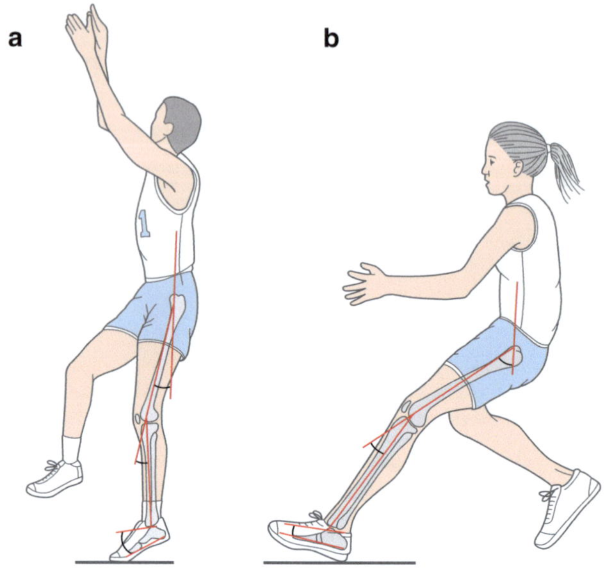

Fig. 3.1 Landing posture(**a**) Safe landing posture(**b**) Unsafe landing posture that may cause ACL injury (Adapted from Ref. [1])

3.2 Epidemiology

The incidence is quite high: 68.6 per 100,000 persons per year in the state of Minnesota, USA. Males are injured about 1.5 times more frequently than females [2].

However, in certain sports, women were injured more frequently than men, 3.5 times in basketball, 2.7 times in soccer, and 1.2 times in lacrosse. On the other hand, in sports such as skiing, there is no difference in injury incidence between men and women [3].

POINT

In sports involving cutting movements, women are injured more frequently.

References

1. Boden BP, et al. Video analysis of anterior cruciate ligament injury: abnormalities in hip and ankle kinematics. Am J Sports Med. 2009;37:252–9.
2. Sanders TL, et al. Incidence of anterior cruciate ligament tears and reconstruction: a 21-year population-based study. Am J Sports Med. 2016;44:1502–7.
3. Prodromos CC, et al. A meta-analysis of the incidence of anterior cruciate ligament tears as a function of gender, sport, and a knee injury-reduction. Arthroscopy. 2007;23:1320–5.

Chapter 4
Symptoms, Natural History After ACL Injury

4.1 Subjective Symptoms

1. Acute phase

At the moment of ACL injury, an anterior dislocation of the tibial plateau – reduction has occurred. Therefore, patients often describe that they felt the knee was dislocated. The patient then complains of collapse and weakness. Swelling due to intra-articular hematoma/hemarthrosis often follows.

2. Chronic phase

The patient often suffers from repetitive giving way due to anterior subluxations or subjective instability during activities. With time, the patient may develop a catching sensation or locking symptoms due to subsequent meniscus tears. It is not uncommon that patients complain of difficulty walking with a locked knee.

4.2 Natural History

Usually, the swelling disappears within a few weeks, moving into the chronic phase. In cases with no /minor meniscus or articular cartilage damage, the patient is almost asymptomatic in activities of daily living.

However, when the patients return to sports that involve jumping, cutting, and sudden stops (basketball, handball, judo, football, etc.), they suffer from giving way or instability.

The natural course of the untreated patients are deteriorated over time: (1) repetitive giving ways; (2) meniscus, articular cartilage damage; (3) secondary osteoarthritis [1]. The deterioration depends on body weight, level of sports activity, gender,

K. Shino, *Essence of Anterior Cruciate Ligament*,
https://doi.org/10.1007/978-981-99-6536-6_4

anatomical characteristics (e.g., posterior inclination angle of the tibial plateau), and other factors (Fig. 4.2).

In the case of inappropriate (reconstructive) surgery (see Chap. 11), which will be discussed later, this path to secondary osteoarthritis is accelerated.

POINT

Patients who are able to perform sports activities involving cutting movements without subjective instability, even though they may be unstable objectively, are called "copers". On the other hand, patients who experience instability or giving way even during light sports activities are called "non-copers".

However, in my opinion, there are no patients who can be strictly called "coper". There were many cases of athletes who had refused treatment and returned to work or sports, but they later suffered from repeated giving way of the knees and developed meniscus injuries within a few years, and came to us seeking for treatment. Such cases were initially considered to be coper, but in fact, they should not be considered coper (Fig. 4.1). We should simply assume that "substantial lesions lead to functional impairment" and simply treat the lesions (Fig. 4.2).

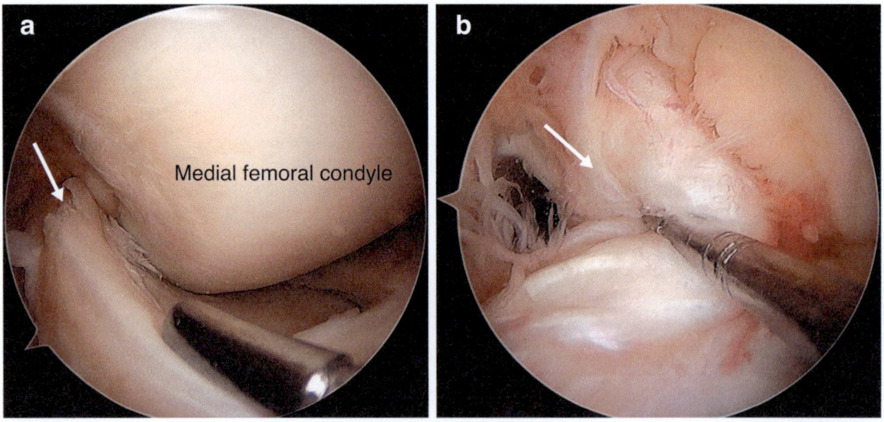

Fig. 4.1 Arthroscopic view of a 41-year-old man (martial artist) suffering from locking due to medial meniscus injury in the right knee
(**a**) The medial meniscus shows a bucket-handle tear, which is displaced to the intercondylar notch (arrow)
(**b**) The ACL (arrow), although continuous, was clearly lax and exhibited obvious anterior instability objectively
The patient did not remember when he injured his ACL and had never felt any instability. Although he could be called a "coper," he developed the meniscus injury secondary to ACL injury. Therefore, he cannot be called a coper in a strict sense.

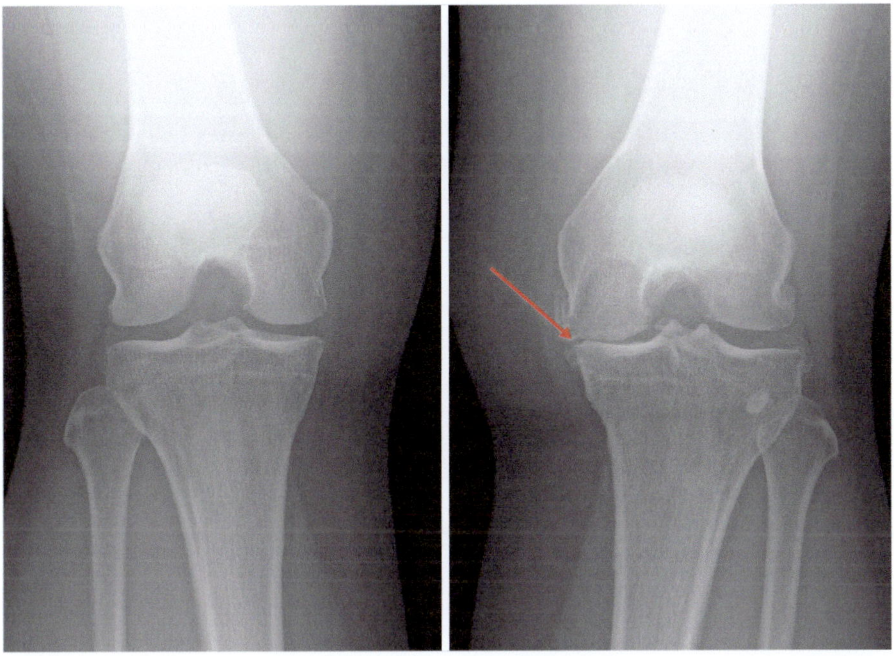

Fig. 4.2 Antero-posterior X-ray image of a 37-year-old woman suffering from walking difficulty due to the locked left knee due to the torn medial meniscus. She sustained the ACL injury 20 years ago but had been left untreated
The left knee showed typical arthritic changes with marked narrowing of the medial joint space and osteophyte formation (arrow). While this patient had been refraining from sports activities for 20 years, a highly active patient could fall into this stage within a few years

Reference

1. Noyes FR, et al. The symptomatic anterior cruciate-deficient knee. Part I: the long-term functional disability in athletically active individuals. J Bone Joint Surg Am. 1983;65:154–62.

Chapter 5
Diagnosis of ACL Injury

5.1 History Taking

If the pathomechanism of ACL injury is taken in mind, history taking could make it almost possible to diagnose the injury without physical examination or imaging. The followings are checkpoints when taking a history at the time of initial injury, in the acute phase, or in the chronic phase.

1. Checkpoints at the time of initial injury/acute phase

   *Types of sports: Basketball, handball, soccer, apparatus gymnastics with cutting and sudden stops, contact sports such as rugby football or judo.
   *Position at the time of injury: Posterior inclination of the trunk, Extended knee, Knee abduction.
   *Audible pop at the time of injury
   *Sense of dislocation/subluxation at the time of injury
   *Sense of collapse immediately after the injury: Spontaneous reduction after dislocation results in inability to continue sports due to weakness. Some cases with bone bruise, articular cartilage or meniscus injuries, may become unable to walk (Fig. 5.1).
   *Joint swelling within 12 h of injury = hemarthrosis (positive in about 90% of cases)

2. Symptoms in chronic phase

Patients often complaints of giving way or buckling of the knee, but it is not uncommon for them to complain of catching sensation, or locking due to meniscus inju ries. The patient's history at the time of initial injury should be noted (Fig. 4.1).

K. Shino, *Essence of Anterior Cruciate Ligament*,
https://doi.org/10.1007/978-981-99-6536-6_5

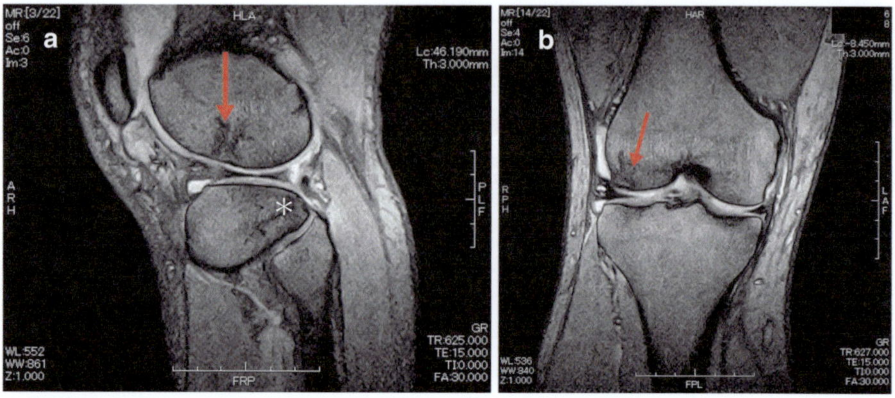

Fig. 5.1 MRI findings of bone bruise in ACL-injured knee (**a**: sagittal section, **b**: coronal section) It frequently occurs on the lateral femoral condyle (arrow) and the posterior margin of the lateral tibial plateau (*). It is like an insufficiency fracture, sometimes leading to a minor depression of the articular surface. Association of this lesion often makes it difficult to walk with weight bearing

5.2 Evaluation of Instability

If an ACL injury is suspected during the history taking, manual laxity examination follows.

1. **Manual laxity examination**

The test is performed in a non-weight bearing position by manually applying force to induce an anterior (lateral) subluxation or to see if the subluxation is reduced. The patients often feel apprehension leading to the thigh muscle contraction if the knee is unstable due to ACL injury. Thus it should be kept in mind that false-negative results are not uncommon in the examination. Therefore, accurate findings/quantitative evaluation can be obtained only under anesthesia without muscular defense.

Also, keep in mind that this is only a test under manual/*in vitro* force at a load level of about 200 N at most.

Pay attention to the amount of displacement (0~5 mm: +, 5~10 mm: ++, 10~15 mm: +++, 15 mm~: ++++), the quality of the end point (hard or soft), and the pateints' apprehension (due to induced subluxation).

In addition, physiological joint laxity varies greatly from person to person, so it is important to compare with the healthy side.

(1) Lachman test/anterior drawer test in extension [1]

It is an anterior drawer test performed in the supine position with the knee flexed 15–30°.

This test is essential to detect instability due to ACL injury.

The distal thigh is grasped just above the patella from the lateral side, and the proximal end of the calf is grasped from the medial side. Then the tibia is pulled anteriorly. Attention should be paid not only to the displacement of the tibia, but

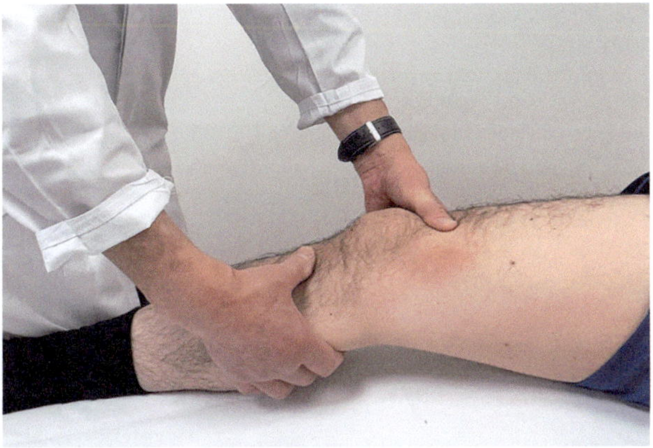

Fig. 5.2 Lachman test: Anterior drawer test with the knee slightly flexed

also to the quality of the end point. The normal knee shows firm end point, while the ACL-injured knee, soft end point. Fig. 5.2).

(2) Anterolateral subluxation test

1) Jerk test (N-test in Japan): In the supine position, the knee is flexed 60° with the tibia internally rotated and the knee is abducted. Then the knee is extended, pushing the fibular head anteriorly with the examiner's thumb. If the test is positive, a sudden anterior subluxation (click) is felt around 30 to 15°. As the patient feels apprehension at the time of the subluxation, it is important to perform the test as quickly as possible to avoid muscular defense leading to false-positive result.

While this test has the advantage of reproducing the subluxation, the subject is prone to defensively contract thigh muscles, leading to false-negative result. Therefore, there is a discrepancy in the results between awake and under anesthesia. When performed in awake condition, patients' apprehension may suggest a positive result.

2) Pivot shift: In the supine position, the tibia is internally rotated and the knee is abducted at 10° flexion. Then the knee is flexed while the calf is held by the examiner's hand. If positive, a click is detected at 30–40° of knee flexion along with reduction of anteriorly subluxed lateral tibial plateau. This test is a reversed version of the Jerk test.
3) Flexion-rotation drawer test [2]: In my opinion, it is easier to get relaxation than the other two tests above. The knee is flexed 20°. Lift up the lateral tibial plateau while applying abduction to the knee (the lateral tibial plateau is displaced forward, while the lateral femoral condyle is displaced backward by gravity. The lateral tibial plateau is subluxed anteriorly). Next, flex the knee gently with mild lateral rotation of the tibia, then the subluxed lateral tibial plateau is reduced by the pulling of the tense iliotibial band (Fig. 5.3).

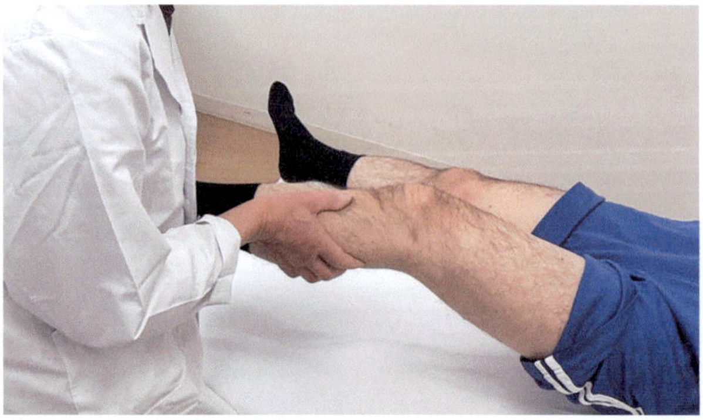

Fig. 5.3 Flexion-rotation drawer test
Lift up the proximal calf from the lateral side. Then, the subluxed lateral tibial plateau is reduced by flexing the knee from 20° with abduction and slight external rotation of the tibia

(Author's own experience)

When I was a resident, I was taught how to perform Jerk test by the famous knee surgeon in front of his patient. After his demonstration showing positive result, I tried to mimic the test for several times according to his guidance. However, no clicks or subluxations were detected. Several years later, a patient performed the subluxation – reduction by himself on his bed, and I finally understood tips of the test (Fig. 1.1). This suggests how important it is to perform the test quickly to avoid the effect of muscular defense of the patient. In other words, we need to be careful because false negative results are often obtained due to the apprehension in the examinee.

POINTS

1. Knee flexion angle

The classic anterior drawer test performed in 90° flexion has a high rate of false-negative results, thus it is useless for detecting instability due to ACL injury.

2. Manual laxity examination using manual load in awake condition

Manual laxity examinations are performed at low load level up to 200 N, to induce subluxation-reduction phenomen. Because subluxation is accompanied by apprehension in the examinee, muscular defenses are inevitable. Therefore, they easily show false negative results. With this in mind, one should not interpret a negative/minimal result as normal/no injury.

3. Rotatory instability

The femoral attachment of the ACL is located posterior to the medial wall of the lateral condyle and the tibial attachment is located slightly medial to the anterior tibial plateau. Thus the ACL runs in a posterior-lateral to anterior-medial direction. Therefore, when the ACL is torn or damaged, the knee shows an anterolateral subluxation/anterolateral instability. In the anterolateral subluxation tests described above, internal tibial rotation is applied only for inducing anterolateral subluxation.

The term rotatory instability was proposed by Hughston JC [3], who stated that it was important to control rotatory instability by tenodesis or capsulorrhaphy, leaving the torn ACL. However, the history has shown that his statement was completely wrong. The term rotatory instability should not be used so readily.

4. Partial ACL injury

When manual laxity examination shows mild instability due to some continuity of the ligament, the diagnosis of "partial injury" may be made. However, even areas of continuity that appear normal on arthroscopy may not be microscopically normal and may show internal damage [4]. There does not seem to be a "partial injury" in which only a part of the ligament is damaged and the rest is normal.

5. Abduction stress test for the ACL-injured knee

ACL is often injured with abduction, and it is not uncommon to find palpable medial instability on abduction stress test due to concomitant medial collateral ligament injury (MCL).

However, at the time of performing abduction stress test on ACL-injured knee without MCL injury, the convex lateral tibial plateau is in contact with the convex lateral femoral condyle, compressed and pushed forward resulting in mild anterior subluxation of the tibia. Thus this makes the medial collateral ligament horizontalized. When medial instability is evaluated with the abduction stress test in this condition, the medial joint space may be opened, leading to over-diagnosis of "positive medial instability/medial collateral ligament injury" (in fact, many cases of ACL injuries are incorrectly diagnosed as medial collateral ligament injuries). When the tibial plateau is controlled to prevent the anterior subluxation at the time of performing abduction stress test, this false positive medial instability could be rarely induced [5] (Fig. 5.4). In other words, it is important to keep the femur-tibia positional relationship normal, when performing abduction/adduction stress test.

2. **Quantitative laxity measurement**

(1) Quantitative anterior drawer test (Lachman) using an instrument [1]

The KT-1000 (Fig. 5.5), KT-2000 (MED-Metric), and Knee Laxity Tester (OSI) were commercially available, but their production has been discontinued. The KS Measure (KSM-100) manufactured by Sigmax Japan is now commercially available and easier to use (Fig. 5.6).

The probability of correctly diagnosing ACL injury by measuring anterior instability using these instruments is approximately 80%, which is less than that of a

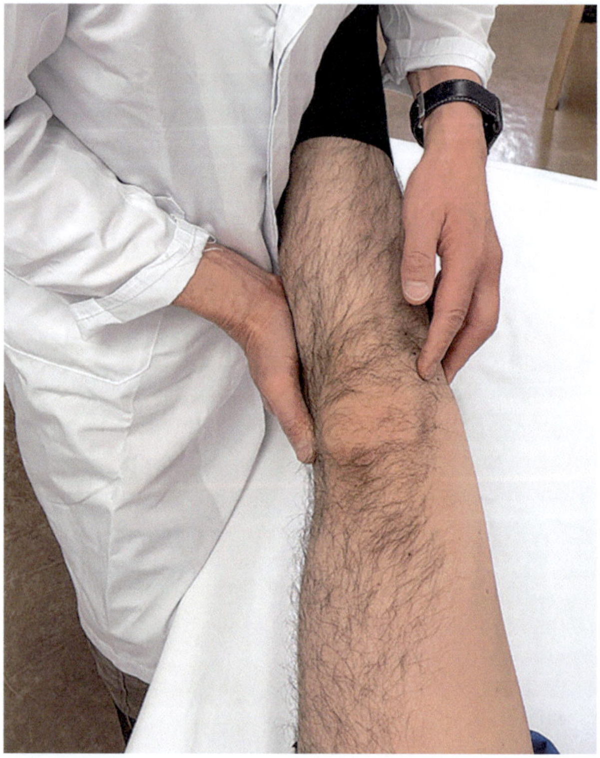

Fig. 5.4 Abduction stress test
Palpate medial joint space opening to detect instability due to medial collateral ligament injury while abducting the knee

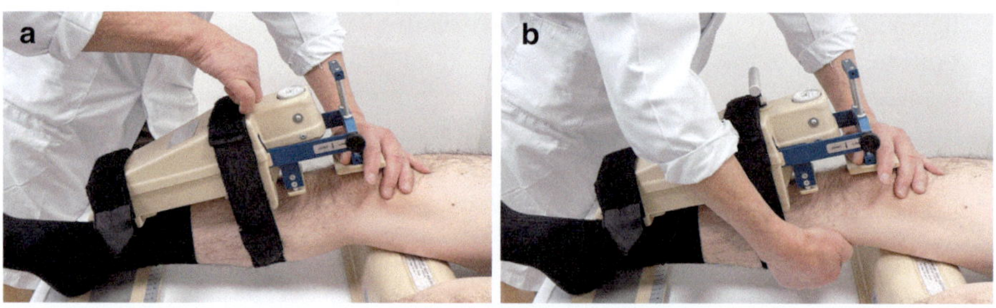

Fig. 5.5 Quantitative anterior laxity measurement using KT-1000
(**a**) In the beginning, 20 lbs/89 N anterior drawer force was applied via a strap to detect a difference in laxity between the healthy and the injured side
(**b**) Currently, the difference between the healthy and the injured side by maximum manual force is used as the KT value

properly performed manual laxity test. However, because of importance of quantifying instability, this method is widely used not only for preoperative evaluation of instability, but also for postoperative evaluation of restored stability after treatment of ACL injury.

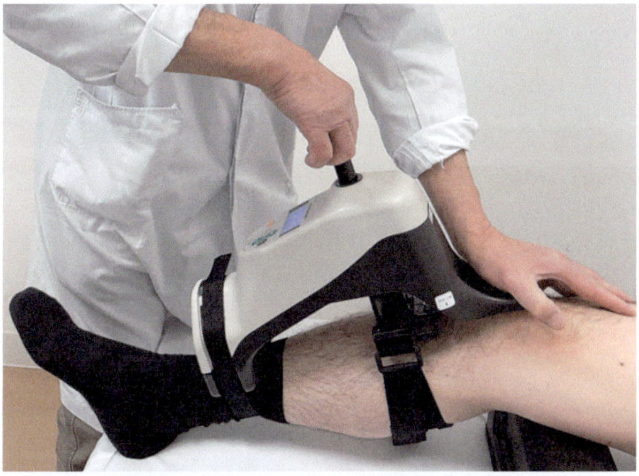

Fig. 5.6 Quantitative anterior laxity measurement using KS Measure (KSM-100) Anterior drawer force can be applied up to 150 N via the strap. A digital display function is provided, and an optional apparatus makes it possible to draw a load-anterior translation curve

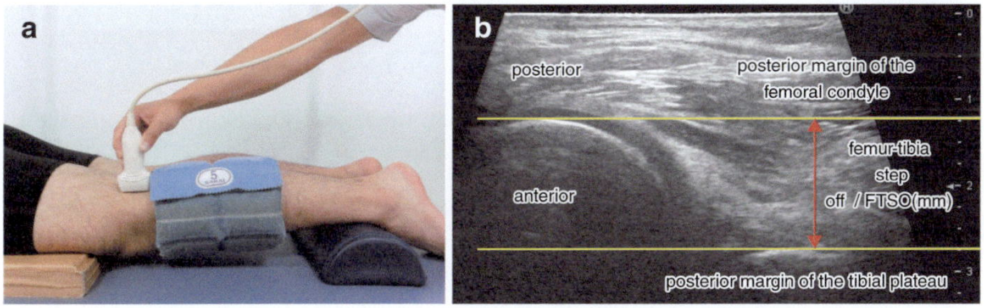

Fig. 5.7 Measurement of anterior tibial displacement using ultrasound imaging system
(**a**) The patient is placed in a prone position with the knee flexed 15°, and a 5-kg weight is placed 10 cm distal to the knee joint line
(**b**) Measure the femur-tibia step off/FTSO in the medial and lateral sides. The mean difference of FTSO between the injured and the uninjured side was 3.6 mm in the medial, and 3.9 mm in the lateral side

(2) Measurement of anterior tibial displacement using ultrasound imaging system

In the prone position, a 5-kg weight is applied to draw the tibia forward, and the femur-tibia step off (FTSO) is measured from behind with an ultrasound probe (Fig. 5.7) [6, 7].

Similar measurements are possible with radiography [8], MRI, and CT, but this method has advantageous in simplicity and in X ray exposure.

POINT
Significance of quantitative laxity measurement

Quantitative laxity measurements, like manual laxity examinations, only measure anterior tibial displacement at low load level up to about 200 N at most/*in vitro* level. It should be noted that even a 2 mm difference in laxity between sides *in vitro* level may be amplified several times greater during *in vivo* activities including jumping. Therefore, the diagnostic accuracy is not so high.

Historically, there have been a great number of reports by subjective evaluation of (well-known) surgeons in which "good" clinical results with excellent restored stability were achieved. However, most of them were not so good as described. In order to avoid this kind of confusion, quantitative laxity measurements are now being used to assess postoperative stability as a common scale of stability, adding objectivity in reported outcomes.

(3) Subjective/functional instability evaluation under weight bearing: trunk backward tilt test [9]

A posture in which the trunk is tilted backward is known as a dangerous position for ACL injury. Therefore, the trunk backward tilt test was devised to evaluate subjective/functional instability of the ACL injury under weight bearing.

In a one-leg standing posture with suppressed forward tilt of the tibia, the subject is instructed to tilt the trunk backward to the maximum extent, and the backward tilt angle is measured (Fig. 5.8). This is a static functional assessment test performed under weight bearing without risk of re-injury.

5.3 Imaging

Since ultrasound is less useful because of location of the ACL, MRI is commonly used. MRI is not only useful for evaluation of the injured ligament itself, but also for evaluation of concomitant injuries such as meniscus and subchondral bone injuries. Thus it is essential for comprehensive evaluation of ACL injuries (Figs. 5.1, 5.9, and 5.10).

In the MRI of the ACL itself, we focus on continuity, running orientation, tension, thickness, and signal intensity (Figs. 5.11 and 5.12). Furthermore, there are cases in which the ligament is elongated and continuous without rupture. In such cases, the ACL itself may appear normal, but the tibia may be displaced several millimeters anteriorly, leading to a diagnosis of ACL insufficiency (Figs. 5.13, 5.14, and 5.15).

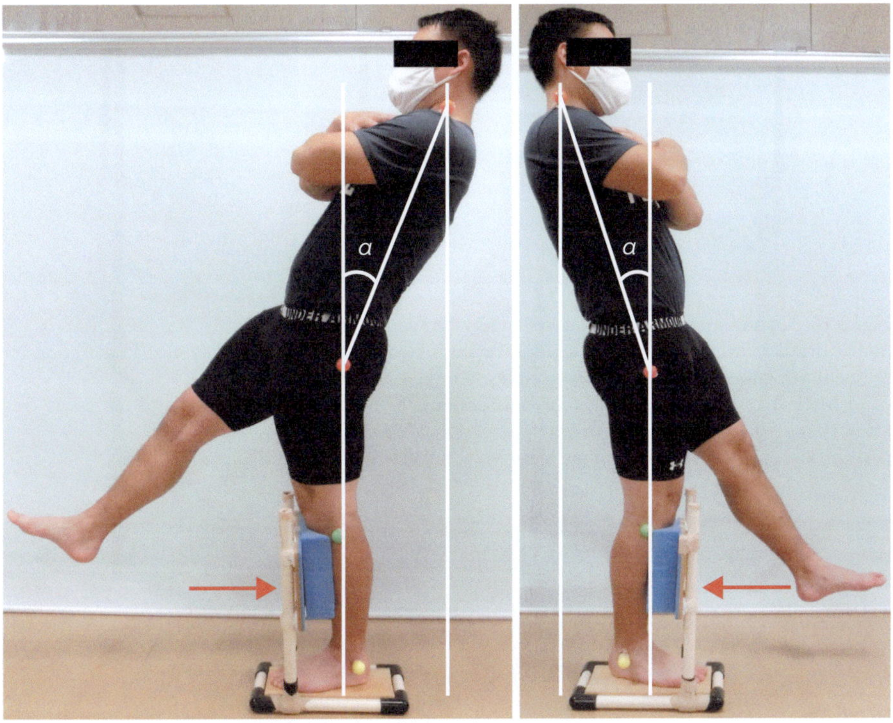

Fig. 5.8 Trunk backward tilt test
The patient stands on one leg, suppressing the anterior tilt of the tibia (→), and tilts the trunk backward. The angle α of trunk backward tilt is measured and evaluated by the difference between sides. The average of α is 17° on the uninjured side and 8° on the injured side

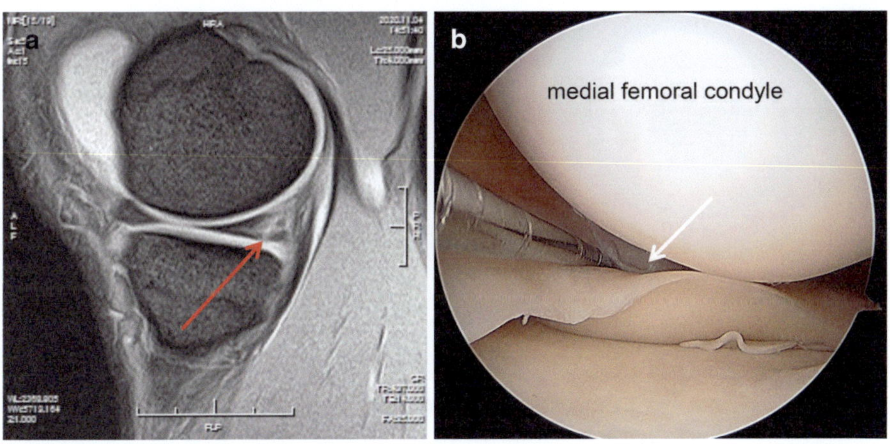

Fig. 5.9 Medial meniscus injury associated with ACL injury
(**a**) MRI: T2*-weighted image (sagittal section): Note high signal intensity area (red arrow) in the posterior segment.
(**b**) Arthroscopic view: longitudinal tear in posterior segment (white arrow)

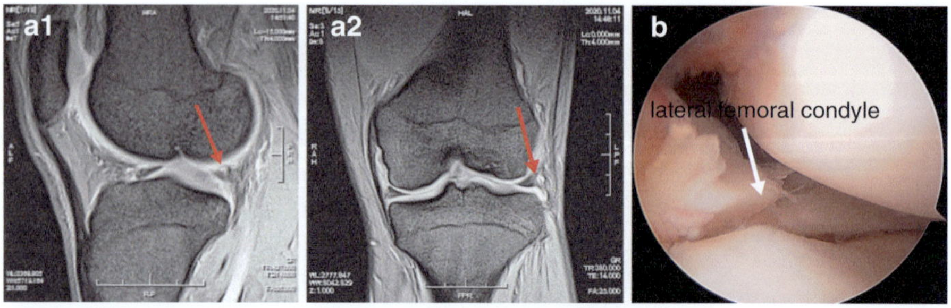

Fig. 5.10 Lateral meniscus injury associated with ACL injury
(**a**-1) MRI: T2*-weighted image (sagittal section): Note high signal intensity area in the posterior segment, irregular meniscus shadow (red arrow)
(**a**-2) MRI: T2*-weighted image (coronal section): Note extrusion of the lateral meniscus (red arrow) suggesting disruption of the hoop structure of the meniscus
(**b**) Arthroscopic view: Note oblique tear (white arrow) in the posterior segment/horn

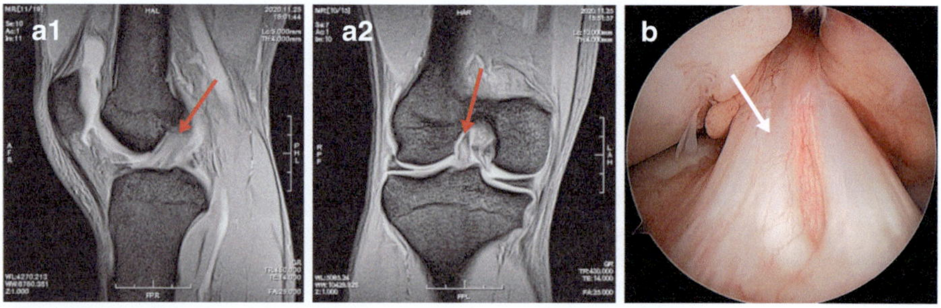

Fig. 5.11 Normal ACL: right knee
(**a**-1) MRI: T2*-weighted image (sagittal section); (**a**-2) (coronal section). In both images, tense fibrous structure between the femur and the tibia is seen (red arrow)
(**b**) Arthroscopic view. A tense fibrous structure is seen, covered by a thin synovial membrane (white arrow)

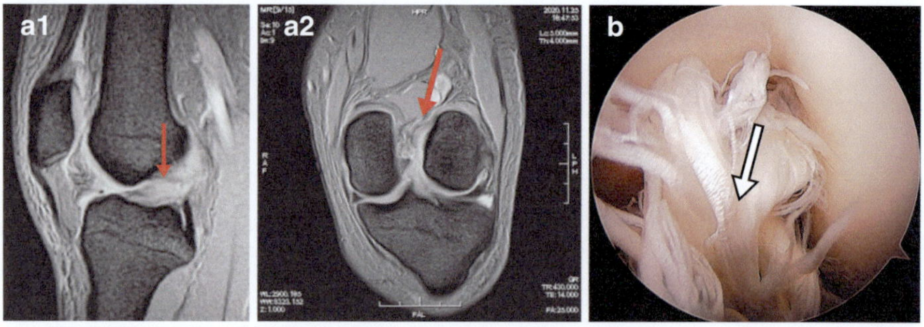

Fig. 5.12 Injured ACL/Acute case (left knee)
(**a**-1) MRI: T2*-weighted image (oblique sagittal section), (**a**-2) (oblique coronal section). In both images, there is no continuity of the tense ligament, which is wavy with inconsistent signal intensity (red arrows). There is a subtle anterior subluxation of the tibia
(**b**) Arthroscopic view of the injured ligament showing a mop-end tear (white arrow)

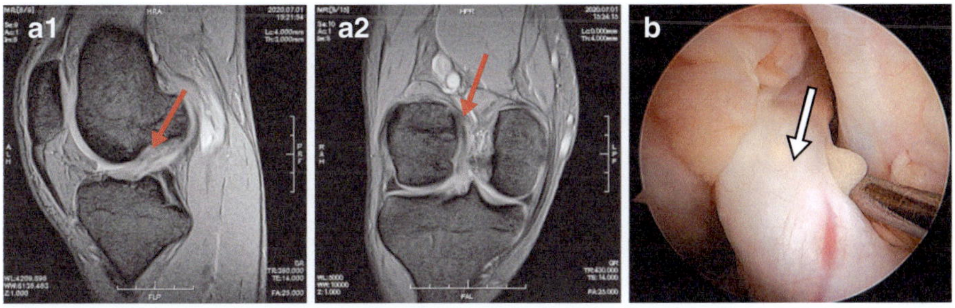

Fig. 5.13 Injured ACL/Chronic case (left knee)
(**a**-1) MRI: T2*-weighted image (oblique sagittal section); (**a**-2) (oblique coronal section). Although continuity of the ligamentous structure can be seen, tension is insufficient, the ligament is wavy with inconsistent signal intensity (red arrows)
(**b**) Arthroscopic view of the injured ligament showing thin and lax appearance (white arrow)

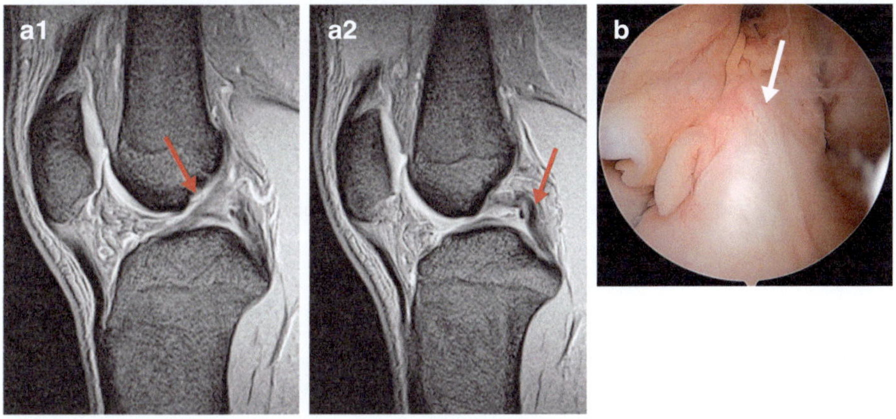

Fig. 5.14 Another example of injured ACL/Chronic case (left knee)
(**a-1**) MRI: T2*-weighted image (sagittal section): The ACL is continuous, but thin and lax (red arrow). Anterior tibial subluxation of several millimeters is noted
(**a-2**) MRI: T2*-weighted image (sagittal section): Buckling of the posterior cruciate ligament (red arrow) is noted due to anteriory-displaced tibia due to ACL insufficiency
(**b**) Arthroscopic view: The remnant tissue is slightly thin and loose with hypervascularity (white arrow)

One of the most common fractures associated with ACL injury is an avulsion fracture of the lateral border of the tibial plateau where the lateral capsule attaches (Segond fracture). This fracture is assumed as a result of an anterolateral dislocation of the tibial plateau, suggesting ACL injury (Fig. 5.16).

POINT
Femur-tibia positional relationship

Even if the ACL itself appears normal, attention should be paid to the femur-tibia positional relationship. If the ligament is elongated, mild anterior tibial subluxation should be noted.

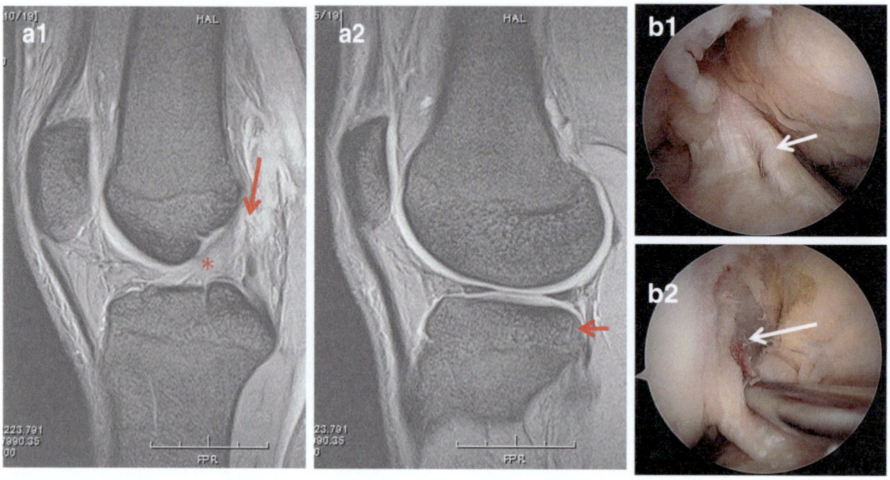

Fig. 5.15 An example to diagnose as an injured/incompetent ACL with difficulty (right knee)
(**a**-1) MRI T2*-weighted image (sagittal section): ACL (*) shows continuity with good tension. High signal intensity area (red arrow) is seen near the femoral attachment site
(**a**-2) MRI T2*-weighted image (sagittal section): Minor anterior subluxation of lateral tibial plateau is noted (red arrow)
(**b**-1) Arthroscopic view of the ACL. Note continuity with abnormal laxity(white arrow)
(**b**-2) Arthroscopic view of the ACL. Hemorrhagic scar near the femoral attachment site (white arrow)

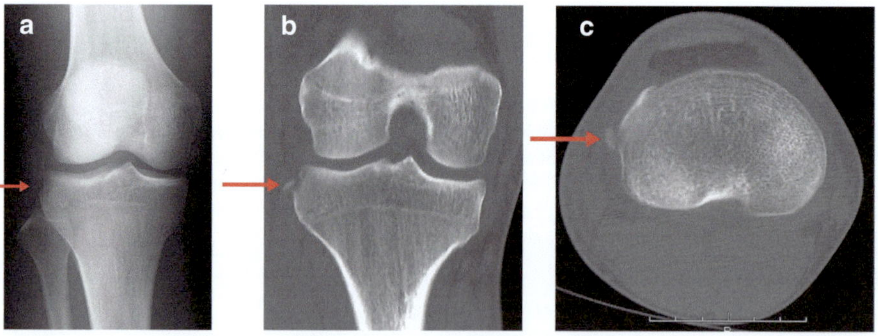

Fig. 5.16 Avulsion fracture of tibial attachment of lateral articular capsule/Segond fracture associated with ACL injury (red arrow) (right knee)
This rare fracture is highly suggestive of ACL injury
(**a**) Plain X-ray, A-P view, (**b**) CT frontal section, (**c**) CT axial image

5.4 Arthroscopic Diagnosis

Although this technique can be performed under local anesthesia, it is invasive and should not be used as a diagnostic method for ACL injury. It should be performed as the first step in an arthroscopic surgical procedure to confirm ACL injury and to evaluate concomitant injuries of the intra-articular structures.

5.5 Differential Diagnosis

Those with recurrent dislocation of the patella also suffer from giving way/instability of the knee. It can be differentiated by physical examination (apprehension sign of the patella/Fairbank sign: lateral patellar deviation test in extension), history (often caused by minor trauma), and radiographs (Skyline image). The apprehension test (Fig. 5.17) is particularly important. If there is any confusion, an evaluation of patellofemoral joint congruence in the extension by CT or MRI is necessary.

When the patient presents with vague complaints of knee instability and a positive anterior drawer sign in 90° flexion, false-positive due to posterior cruciate ligament insufficiency may be doubted. This can be easily confirmed by taking a lateral radiograph in the knee upright position [10] (Fig. 5.18).

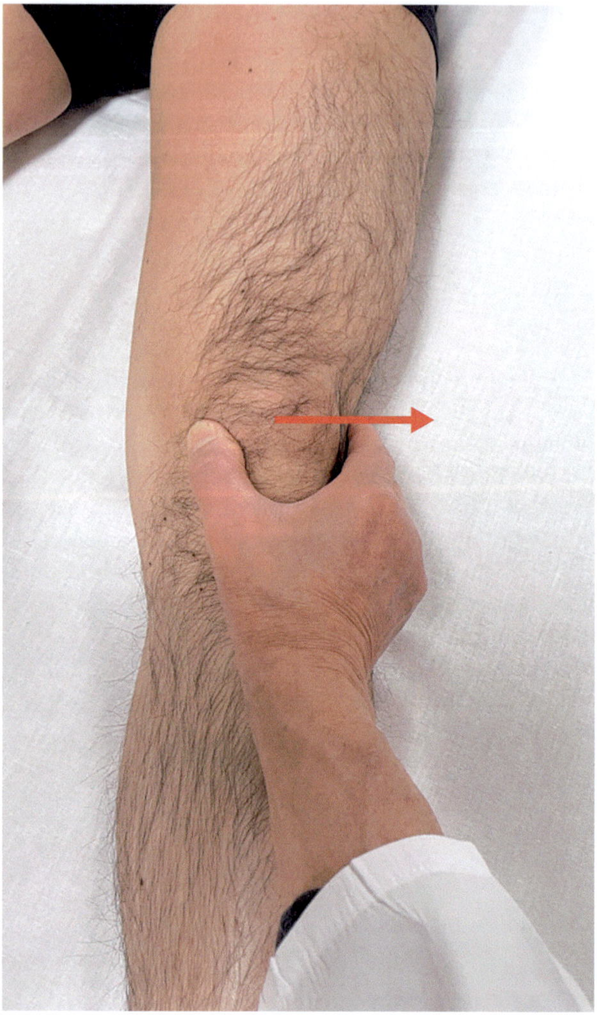

Fig. 5.17 Apprehension test/lateral patellar deviation test/Fairbank test If the test is positive, the subject complains of apprehension when the patella is manually pushed laterally (red arrow)

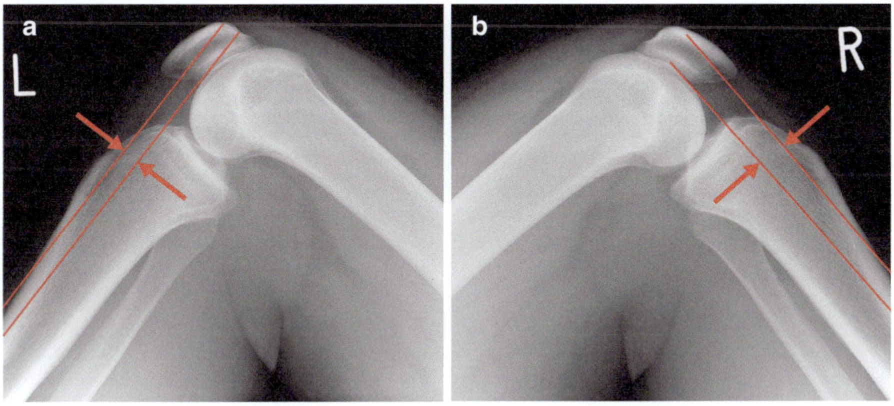

Fig. 5.18 Lateral X-ray image in knee upright position
To avoid overlooking posterior cruciate ligament (PCL) insufficiency, lateral radiographs in knee upright position in 90° flexion is recommended. (**a**) The PCL-injured left knee. (**b**) Uninjured right knee. Note drop back of the tibia due to PCL insufficiency with gravity. Side-to-side difference in the tibia-femur step off (shown on the figures) can be used as a parameter

References

1. Torg JS, et al. Clinical diagnosis of anterior cruciate ligament instability in the athlete. Am J Sports Med. 1976;4:84–93.
2. Noyes FR, et al. Arthroscopy in acute traumatic hemarthrosis of the knee. incidence of anterior cruciate tears and other injuries. J Bone Joint Surg Am. 1980;62:687–95.
3. Hughston JC, et al. Classification of knee ligament instabilities Part I. The medial compartment and cruciate ligaments Part II. J Bone Joint Surg Am. 1976;58(159–172):173–9.
4. Jog AV, et al. Is a partial anterior cruciate ligament tear truly partial? Arthroscopy. 2020;36:1706–13.
5. Ohori T, Shino K, et al. Varus-valgus instability in the anterior cruciate ligament-deficient knee: effect of posterior tibial load. J Exp Orthop. 2017;4:24.
6. Gebhard F, et al. Ultrasound evaluation of gravity induced anterior drawer following anterior ligament lesion. Knee Surg Sports Traumatol Arthrosc. 1999;7:166–72.
7. Matsuo T, Shino K, et al. Accuracy test of anterior knee joint instability assessment by ultrasound. JOSKAS. 2016;41:480–1. (in Japanese)
8. Hiramatsu K, Shino K, et al. Anterior tibial loading on the calf enhances anterior tibial translation in the anterior cruciate ligament deficient knee in the anterior gravity radiographic view. Knee. 2020;27:1764–71.
9. Matsuo T, Shino K, et al. Quantitative evaluation of functional instability due to anterior cruciate ligament deficiency. Orthop J Sports Med. 2020;8:2325967120933885.
10. Shino K, et al. The gravity sag view: a simple radiographic technique to show posterior laxity of the knee. Arthroscopy. 2000;16:670–2.

Chapter 6
Treatment Strategy for ACL Injury

6.1 Background

When the ACL is injured, the healing process takes place to some extent. However, even with conservative treatment such as immobilization with limitation of motion, not enough scar/healing tissue is formed in the intra-articular environment, and only tiny, lax scar tissue with or without some continuity remains (Fig. 5.14). In other words, the knee becomes unstable and follows the natural history described in Chap. 4. If sports activities (including jumping, stopping, cutting, full speed running, etc.) are continued without any effort, giving way (anterior dislocation or subluxation of the tibial plateau → reduction) may occur repeatedly, resulting in meniscus and articular cartilage damage. This leads to secondary osteoarthritis at an early stage. Even if the patient abstains from sports activities, the minor anterior subluxation of the tibial plateau persists, and the long-term outcome is the same as in the case of continuing sports activities (Fig. 4.2).

Immobilization such as plaster casts or bandages do not heal the injured ligament to a stable knee (although there are exceptional cases (Fig. 6.1) in which injured ligament heals spontaneously). Therefore, in principle, conservative treatment aiming at healig the ligament is meaningless.

K. Shino, *Essence of Anterior Cruciate Ligament*,
https://doi.org/10.1007/978-981-99-6536-6_6

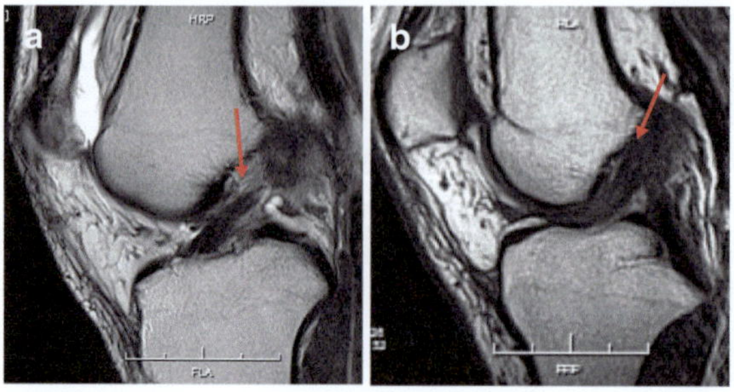

Fig. 6.1 MRI (T1-weighted sagittal section) of a 51-year-old man who sustained ACL injury of the right knee while skiing
(**a**) At 2 days: ACL shows tear in its proximal part (red arrow)
(**b**) After 100 days: The ACL shows spontaneous healing (red arrow). The knee was clinically stable Such spontaneous healing is rare, and the author has experienced only several cases

The only way to achieve the treatment goal of correcting the anterior subluxation of the tibial plateau and restoring to a stable knee without abnormal tibia-femur relationship is through meticulous surgical treatment.

POINT

In case of the medial collateral ligament: extra-articular ligament, when injured, proliferative granulation-like scar/healing tissue is formed. Then the healing tissue is restructured, gets matured, and becomes re-tensioned. Thus the knee restores stability. (Fig. 6.2). In other words, the healing process is similar to that of fracture healing. Therefore, surgery is not necessary for isolated medial collateral ligament injuries, with rare exceptions.

In contrast to the medial collateral ligament, the ACL: intra-articular ligament, when injured, does not seem to undergo the above-mentioned healing process. Thus spontaneous healing could not be expected in the injured ACL.

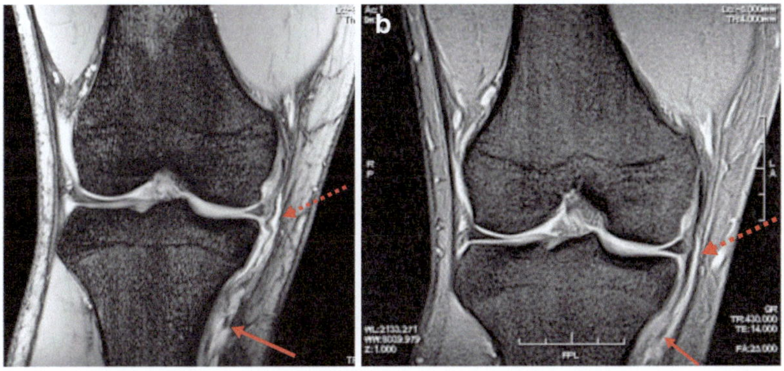

Fig. 6.2 MRI (T2*-weighted coronal section) of an 18-year-old male who sustained the medial collateral ligament injury of the right knee while playing American football
(**a**) At 4 days: The medial collateral ligament is injured in its distal part (solid arrow). The ligament appears lax and wavy (dashed arrow)
(**b**) At 3 months after conservative treatment including soft brace. The distal injured part is slightly thickened and shows healing (solid arrow). The central lax part of the ligament is re-tensioned (dashed arrow). The patient returned to football with the clinically stable knee

6.2 In Acute Phase

First, the concomitant injuries including meniscus tears are evaluated by MRI. The main points of treatment in the acute phase are as follows

1. Not immobilization but range-of-motion exercise to prevent contracture/arthrofibrosis is recommended. Extension exercise is especially important.
2. Arthrocentesis is performed as necessary to drain the hemarthrosis
3. Essentially weight bearing is allowed. However, those with bone bruise may have difficulty in weight bearing.
4. In patients with the medial collateral ligament injuries, range of motion exercise should be performed while controlling abduction with a brace. The injured medial collateral ligament tends to adhere to the surrounding tissues during the healing process, resulting in loss of motion and flexion contracture. Since range of motion exercise does not impair healing of the medial collateral ligament, it should be performed vigorously for several weeks. After the range of motion has been normalized with healing of the medial collateral ligament, ACL reconstruction is performed.
5. Repair of the meniscus should be conducted as early as possible (within 4 weeks after the injury at the latest).
6. There is no urgent need for ACL reconstruction in patients without concomitant meniscus injury. The reconstruction may be scheduled at about 3 weeks after injury, after range of motion is normalized. This minimizes the risk of arthrofibrosis. During this waiting period, some patients may show a sign of spontaneous healing to restore the stable knee due to unknown healing process (see Chap. 9).

6.3 Surgical Procedures

1. Primary repair in acute phase (Fig. 6.3)

As previously mentioned, injured ACLs do not heal spontaneously, and their stumps become atrophic over time. Therefore, some recommend primary repair in the acute phase. However, I do not recommend it for the following reasons:

(1) Acute surgery carries a high risk of complications of arthrofibrosis; (2) If the mop-ends are sutured together, the sutured site is not strong enough to withstand early range-of-motion exercise; (3) In the intra-articular environment, the healing process of proliferative granulation-like healing tissue formation could not be expected as in the medial collateral ligament injuries. Even if it heals to some extent, it becomes a lax ligament-like tissue which could not be robust enough to restore the stability. In fact, unsatisfactory results have been reported [1].

2. Partial ligament reconstruction/ligament augmentation (reconstruction of anterior medial or posterior lateral bundle)

Partial reconstruction, in which only the injured part is reconstructed after arthroscopic observation, has been proposed and good results have been reported

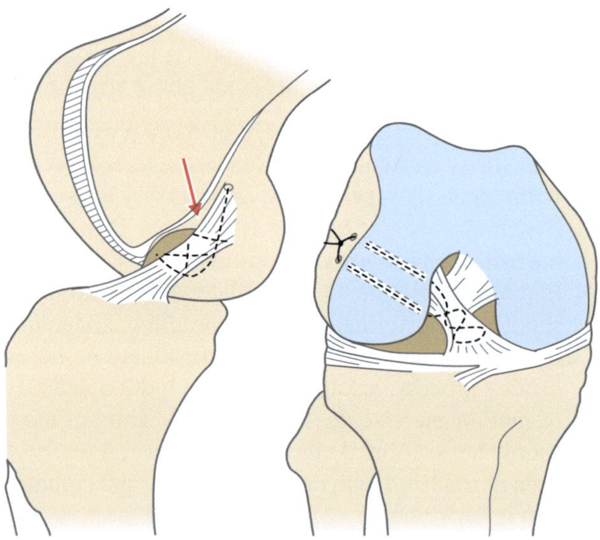

Fig. 6.3 Primary ACL repair performed in the acute phase
The sutures placed in the stump near the femoral attachment site (arrow) are pulled out to the lateral femoral cortex through the small drill holes and tied over the cortex. (Adapted from Ref. [1])

[2]. Since ACL injury is caused by anterior dislocation of the tibia, as mentioned above, it is difficult to discriminate between normal and injured parts. In other words, the normally-looking part is assumed damaged too [3]. The treatment concept should be reconstructing the severely injured part with a graft to stabilize the joint, expecting natural healing of the less damaged part. Since surgical treatment is burdensome for the patient, surgery with the highest success rate should be performed. From this point of view, the author recommends anatomical reconstruction of the entire ligament as much as possible.

3. Entire ligament reconstruction with residual stump preserved

In order to preserve the nerve endings present in the residual stumps of the ligament, some advocate the reconstruction with the stump preserved. As it is essential to create a drill hole at the exact anatomic attachment site, this reconstruction technique is not recommended because of the following reasons: (1) If the drill holes are created at the exact attachment areas, their apertures should be located just under the areas. Therefore, most of the remaining stumps attached to the areas should be sacrificed; (2) The bony surface of the ligament attachment areas cannot be well visualized if the stumps are preserved.

4. Entire ligament reconstruction with the residual stumps excised

It is essential to place the bony tunnel apertures at the exact ligament attachment areas at the time of performing strictly-anatomical ACL reconstruction. Thus bony surfaces of the areas should be well visualized. For this purpose, the soft tissues around the ligament attachment areas are meticulously excised with not mechanical but electro-surgical device to preserve undulation of the bony surface at and near the attachment areas. This makes it possible to delineate the attachment areas for accurate anatomical tunnel creation. This is the author's preferred strategy for anatomical tunnel creation/"bony landmark strategy".

5. Lateral extra-articular procedure

In the twentieth century, Lateral extra-articular reconstruction/tenodesis with the iliotibial band was frequently performed without intra-articular reconstruction [4]. The results were disastrous.

In recent years, there have been some reports of improved outcomes when combined with intra-articular ACL reconstruction (Fig. 6.4) [5]. This additional extra-articular surgery may be unnecessary if the intra-articular ligament reconstruction is performed strictly-anatomically [6].

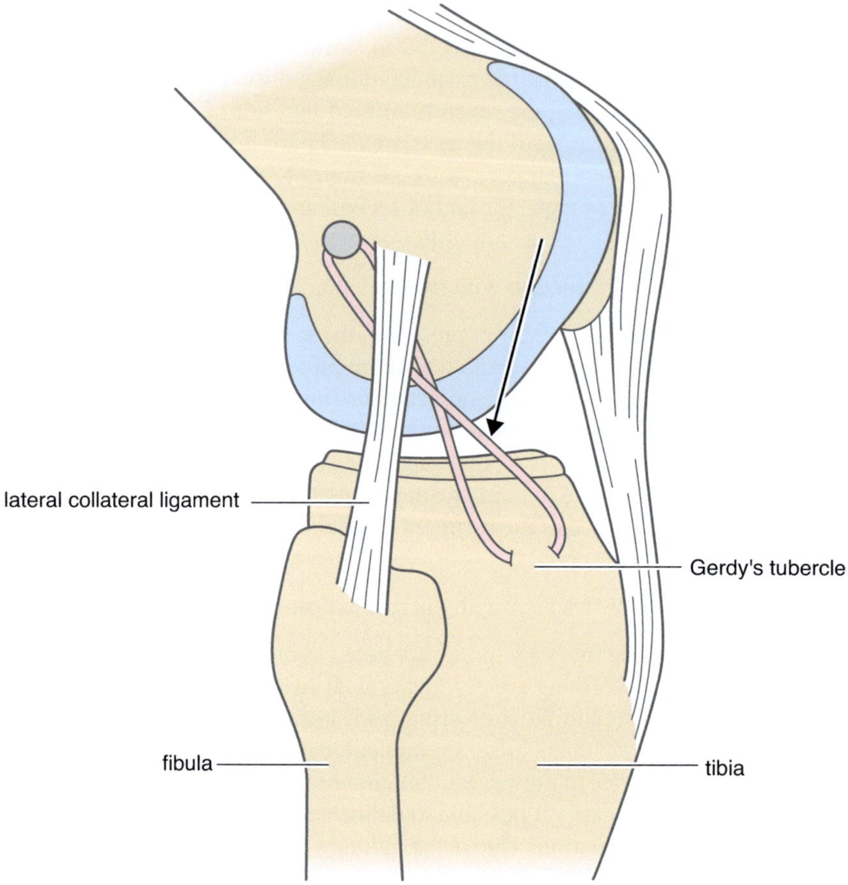

Fig. 6.4 Lateral extra-articular reconstruction/tenodesis (right knee)
The distally-attached iliotibial band or, as in this case, a free tendon graft (e.g., semitendinosus tendon) (arrow) is used
(Adapted from Ref. [5])

References

1. Feagin JA, et al. Isolated tears of the anterior cruciate ligament: a S-year follow-up study. Am J Sports Med. 1976;4:95–100.
2. Nakamae A, et al. Clinical outcomes of second-look arthroscopic evaluation after anterior cruciate ligament augmentation: comparison with single- and double-bundle reconstruction. Bone Joint J. 2014;96-B:1325–32.
3. Jog AV et al. Is a partial anterior cruciate ligament tear truly partial? A clinical, arthroscopic, and histologic investigation. Arthroscopy. 2020;36:1706–13.
4. Losce RE, et al. Anterior subluxation of the lateral tibial plateau. A diagnostic test and operative repair. J Bone Joint Surg Am. 1978;60:1015–30.
5. Magnussen RA, et al. The role of lateral extra-articular augmentation in revision ACL reconstruction. In: Marx RG, editor. Revision ACL reconstruction. New York: Springer; 2014. p. 151–6.
6. Tachibana Y, Shino K, et al. Anatomical rectangular tunnels identified with arthroscopic landmarks result in excellent outcomes in ACL reconstruction with a BTB graft. Knee Surg Sports Traumatol Arhrosc. 2019;27:2680–90.

Chapter 7
Basics in ACL Reconstruction

7.1 Background

The basic treatment of unstable knees due to ACL injury is ACL reconstruction using tendon or the other soft tissue grafts. In the United States, it is estimated that more than 175,000 ACL reconstructions are performed annually. It is one of the seven most frequently-performed orthopedic operative procedures in Japan.

It is needless to say that the goal of ACL reconstruction is to restore stability. However, the stability-restored knee with limitation of range of motion is inferior to the unstable knee in function and prone to secondary osteoarthritis as the knee with residual instability [1].In order to achieve the stability-restored knee without limitation of range of motion, appropriate graft selection and its precise anatomical placement are mandatory. Incorrect or non-anatomical reconstruction results in the knee with residual instability and/or limited range of motion, leading to early secondary osteoarthritis. This kind of improper surgery may be called tenodesis rather than reconstruction.

7.2 Graft Selection

In the beginning, autologous distally-pedicled grafts including iliotibial band or patellar tendon were used. However, because the iliotibial band was not robust enough, or because it was almost impossible to anatomically place the pedicled patellar tendon graft, the procedures with those grafts resulted in poor outcomes.

More recently, the free graft was proved equivalent to the pedicled grafts in their remodeling process, suggesting no need to stick to a pedicled graft [1]. From around 1980, there has been a gradual increase in momentum for the anatomical placement

K. Shino, *Essence of Anterior Cruciate Ligament*,
https://doi.org/10.1007/978-981-99-6536-6_7

of autologous free tendons as graft material. Currently, free autologous tendons are the gold standard, although allogeneic tendons are also used in some areas. Artificial materials such as polyester, PTFE (Teflon), and carbon fibers were used in the past, but they should not be used because of their rapid degradation in the human body.

As for graft selection criteria, it must be a biological material with higher structural properties (higher failure load/stiffness) than the normal ACL [2, 3]. In terms of material properties, it must be higher in tensile strength and elastic modulus. It should be longer than 60 mm, and desirably a bit thinner. The patellar tendon of 8–13 mm width, the multi-stranded hamstring tendon or the quadriceps tendon of 10 mm width are used to mimic the flat, tape-like shape of the native ACL.

Currently-used grafts used for ACL reconstruction are as follows:

1. Patellar tendon autograft with bone plugs (Fig. 7.1a): A patellar tendon with bone plugs of 8–13 mm width and about 15-mm long attached to both ends. The bone plugs and the tendon-bone junction area are advantageous for bone plug-tunnel integration.
2. Multi-stranded hamstring/semitendinosus tendon autograft (Fig. 7.1b): The entire tendon is harvested, then transversely cut into two doubled grafts of 5 to 6 mm in diameter. In case the semitendinosus tendon is not long enough, the doubled gracilis tendon is also used.
3. Quadriceps tendon autograft (with a patellar bone plug) (Fig. 7.1c): A quadriceps tendon of 10-mm width and at least 5-cm length with a patellar bone plug of 15-mm length attached.
4. Allogeneic tendon grafts: Patellar tendon with bone plugs, Achilles tendon, and anterior and posterior tibialis tendons are commercially available in the United States.

In 1981, the author pioneered the use of allograft tendons such as Achilles tendon, anterior/posterior tibialis tendons, and peroneus tendons from amputated limbs, to create an graft of 8–10 mm in diameter by splitting or bundling them [4, 5]. However, the procedure had to be quit because of not poor results but lacking in tissue collecting system. The tissue bank is not yet established in Japan due to social issues.

In some Western countries/regions, such as the United States, cadaver tissue banks have been established, and lyophilized or chemically-treated tendons are available and frequently used (Fig. 7.2).

POINT

The structural properties of the multi-stranded hamstring tendons and the 10-mm wide patellar tendon is much superior to that of the normal ACL. Thus they are recommended for use. However, when the quadriceps femoris tendon is used, it should be noted the graft is 50% thicker than the patellar tendon because of its inferior material properties [3].

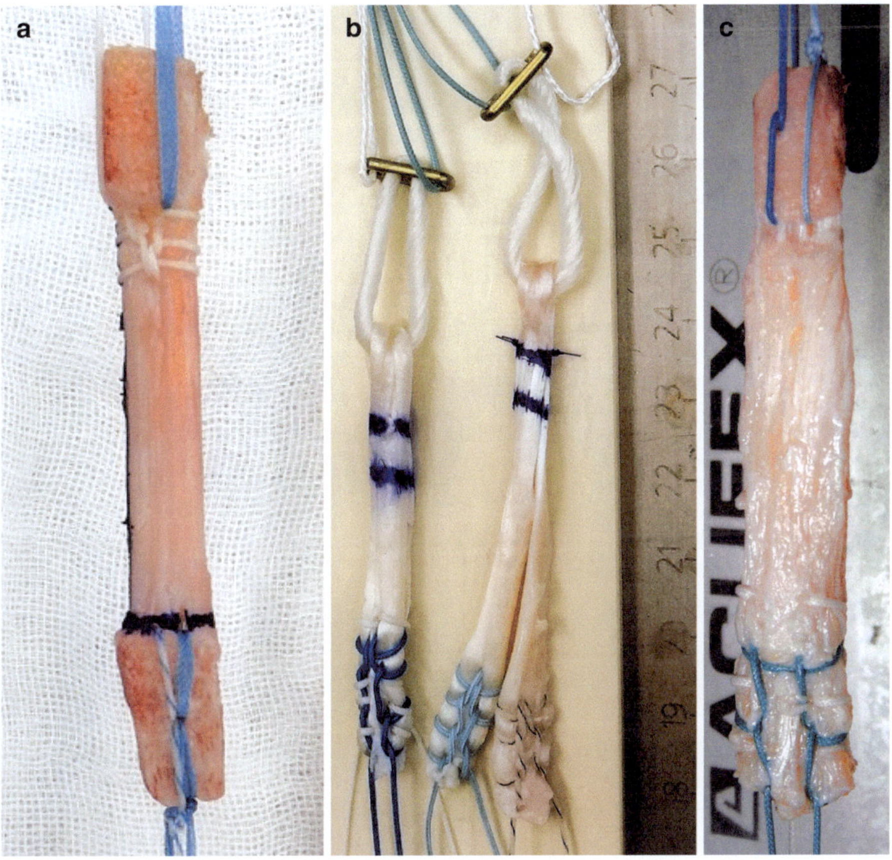

Fig. 7.1 Autologous grafts used for ACL reconstruction
(**a**) Patellar tendon graft with bone plugs: A patellar tendon with bone plugs of 8–13 mm width and about 15-mm length attached to both ends. The bone plugs and the tendon-bone junction area are advantageous for bone plug-tunnel integration
(**b**) Multi-stranded hamstring/semitendinosus tendon graft: The entire tendon is harvested, then transversely cut into two doubled grafts of 5 to 6 mm in diameter. In case the semitendinosus tendon is not long enough, the doubled gracilis tendon is also used
(**c**) Quadriceps tendon graft (with a patellar bone plug): A quadriceps tendon of 10-mm width and at least 5-cm length with a patellar bone plug of 15-mm length attached

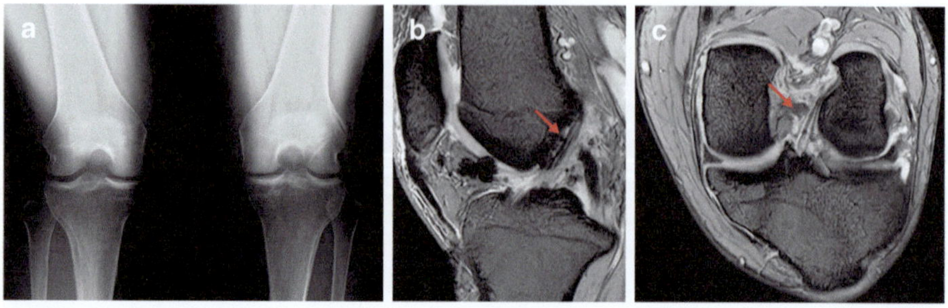

Fig. 7.2 A case with ACL reconstruction with allogeneic tendon graft @ 32 years previously (51 year-old female)
Thirty-two years ago, at the age of 19, she underwent ACL reconstruction in the left knee using fresh-frozen allogeneic tendon without bone plugs.
Recently, a small tear of the posterior horn of the medial meniscus developed.
(**a**) Antero-posterior radiograph: No arthritic changes noted.
(**b**) MRI (T2*-weighted) oblique sagittal slice: The reconstructed ACL (arrow) shows good tension and runs similar to normal ligament.
(**c**) MRI (T2*-weighted) oblique coronal slice: The reconstructed ACL (arrow) shows good tension and runs similar to the normal ligament

7.3 Anatomical Graft Placement

The author believes it mandatory to create tunnel apertures inside the intra-articular ACL attachment areas. This makes it possible to realize an impingement-free graft that does not impinge on the other structures (posterior cruciate ligament, intercondylar notch, etc.) [6].

Therefore, it is necessary to arthroscopically identify the anatomical attachment areas to accurately create anatomical tunnels, which can be consistently achieved by using a radio-frequency device to expose the concave attachment areas while preserving bony undulation around the areas (Fig. 7.3). This “bony landmark strategy” (Fig. 7.3) makes it possible to identify the concave attachment areas [7]. The tibial tunnel is created from the area just medial to the tibial tubercle to the intra-articular tibial attachment area. The tunnel is angled at 45° to the tibial axis. This makes the graft nearly in line with the tibial tunnel in full extension (Fig. 7.4). The femoral tunnel is routed between the lateral femoral cortex (above the lateral femoral epicondyle) and the intra-articular femoral attachment area [8] (Fig. 7.5).

If the intra-articular aperture of the tibial tunnel deviates from the attachment areas, they could not only destroy the insertion of the meniscus [9] (Fig. 7.6) but cause improper orientation of the graft [10].

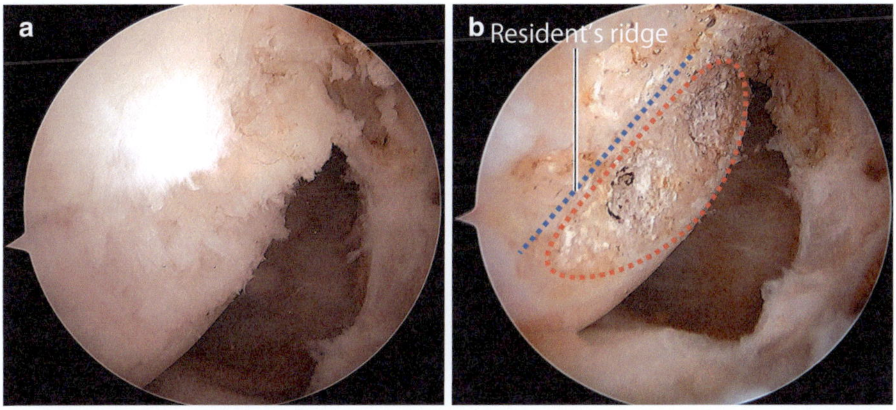

Fig. 7.3 Bony landmark strategy
(**a**) The femoral attachment area viewed through the anterointernal portal. Note identification of the attachment area is not easy.
(**b**) Using a radio-frequency device, the fibrous tissue near the attachment area is removed while preserving undulation of the bony surface. The concave attachment area (crescent-shaped area surrounded by a red dashed line) is exposed, and bony landmarks such as the resident's ridge is identified. This makes it possible to consistently identify the attachment area

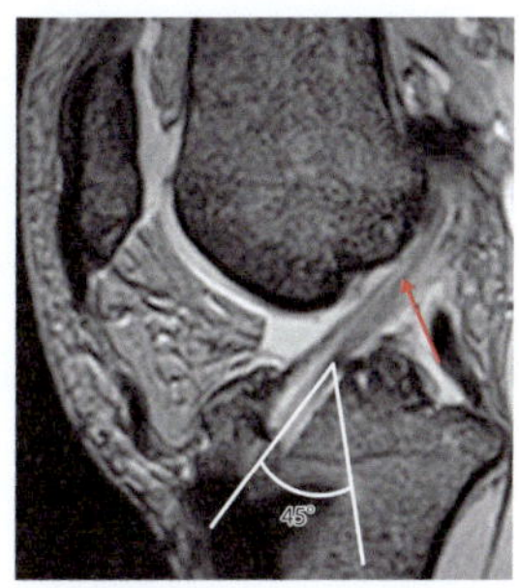

Fig. 7.4 MRI sagittal cross-sectional image (T2*-weighted) of the ACL graft in extension
A properly created tibial tunnel makes a 45° angle with the tibial axis, and is in line with the reconstructed ligament (arrow) in extension

Even if a graft (especially composed of soft tissue only such as hamstring tendon graft) is snugly placed in a just-sized tunnel, it deforms according to the direction of the applied force after graft tensioning at the tibia, because of the graft-bending angle of 70–80° at the femoral tunnel aperture (Fig. 7.7). This deformation leaves a void behind the graft (Fig. 7.8), and results in an anteroinferior deflection of the graft in the femoral tunnel, leading to a more vertically-orientated graft than planned (Fig. 7.9). Obviously, the greater the tunnel size is, the severer this deflection becomes. Therefore, in case of a single-bundle reconstruction using a multi-stranded hamstring tendon graft via a single anatomical femoral tunnel of 8–10 mm, the

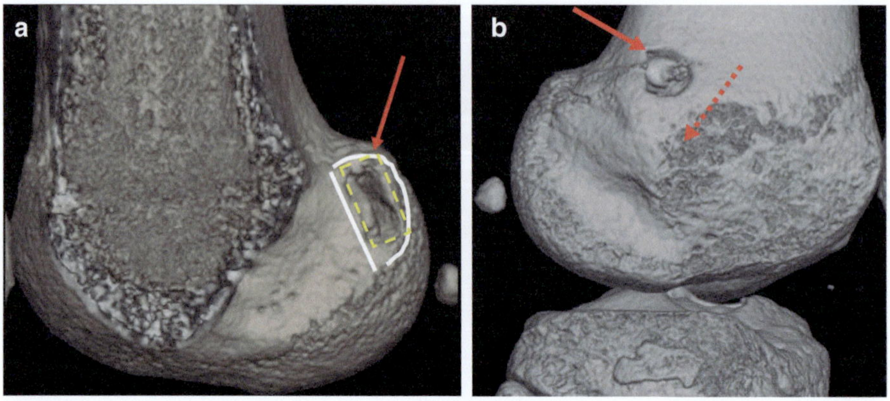

Fig. 7.5 Anatomical femoral tunnel aperture(right knee)
(**a**) 3-D CT image of the lateral wall of the intercondylar notch (right knee). The intra-articular aperture of the anatomical rectangular tunnel (yellow dashed line) is located inside the femoral attachment area of semilunar shape (red arrow)
(**b**) 3-D CT image of lateral femoral cortex showing the tunnel aperture in the lateral femoral cortex (red arrow). The aperture is located just proximal to the lateral epicondyle (dashed arrow)

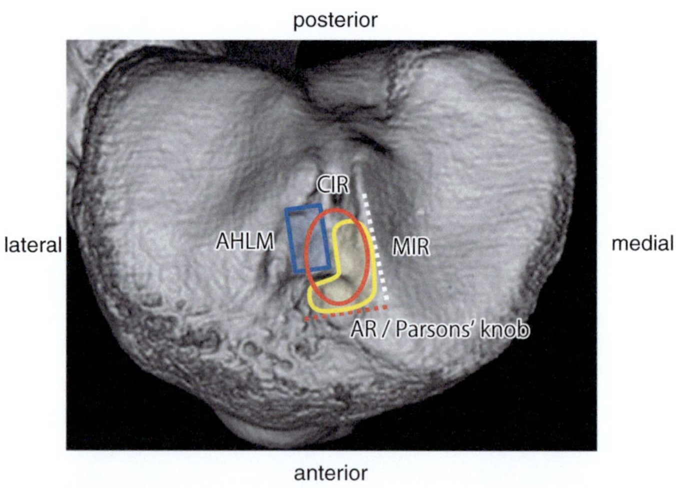

Fig. 7.6 3-D CT image of the tibial plateau (right knee)
ACL attachment area (yellow-lined area) and cylindrical tunnel aperture (red-lined oval area). The boot-shaped ACL attachment area is surrounded by the MIR (medial intercondylar ridge), AR (anterior ridge), and CIR (central intercondylar ridge). The AHLM (anterior horn of the lateral meniscus) is attached to the lateral slope of the CIR.
If a 10-mm diameter tibial tunnel is created to the center of the anatomical attachment area (red-lined oval) from the anterior aspect of the tibia, AHLM is partially injured

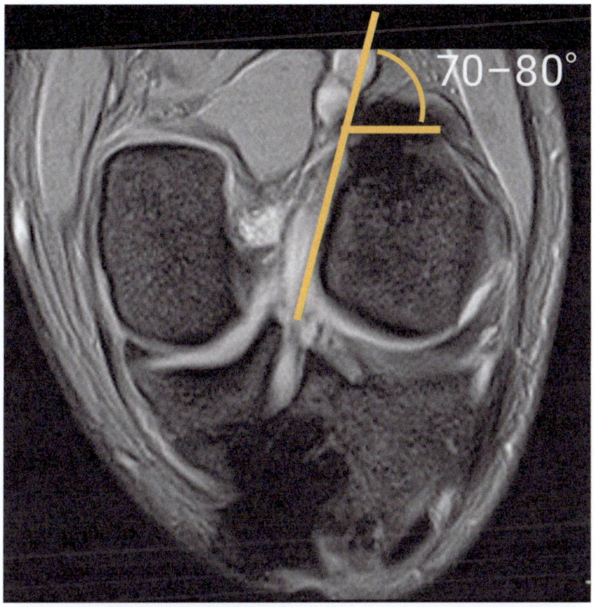

Fig. 7.7 Graft bending angle of 70–80° @ femoral tunnel aperture (MRI coronal section)

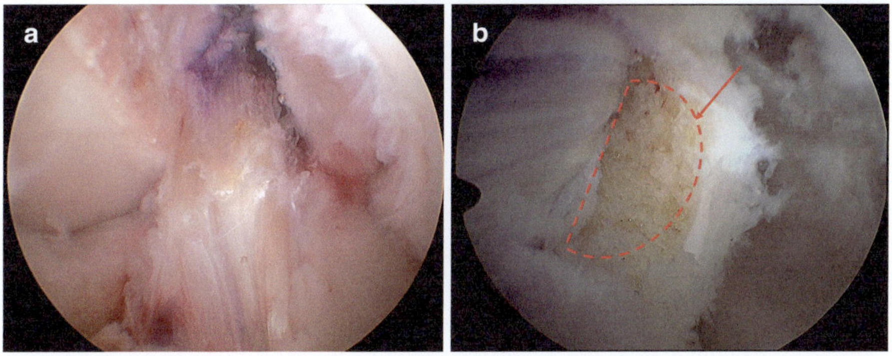

Fig. 7.8 Reconstructed ACL with hamstring tendon graft and femoral tunnel aperture
(**a**) Hamstring tendon graft *in situ* (anterior view)
(**b**) The same graft viewed from behind
Even if a soft tissue graft such as hamstring tendon graft is snugly installed in the just-sized femoral tunnel, it deforms inside the tunnel according to the direction of the applied force after graft tensioning at the tibia, because of the graft-bending angle at the femoral tunnel aperture (Fig. 7.7). This deformation leaves a crescent-shaped void behind the graft (arrow), and results in an antero-inferior deflection of the graft in the femoral tunnel, leading to a more vertically-orientated graft

anterior deviation of the graft inside the tunnel is more prominent. Thus a single bundle reconstruction using multi-stranded hamstring tendons via such a large size femoral tunnel is not recommended. Instead, double or triple bundle reconstruction via multiple smaller size tunnels is recommended in case a hamstring tendon graft is used.

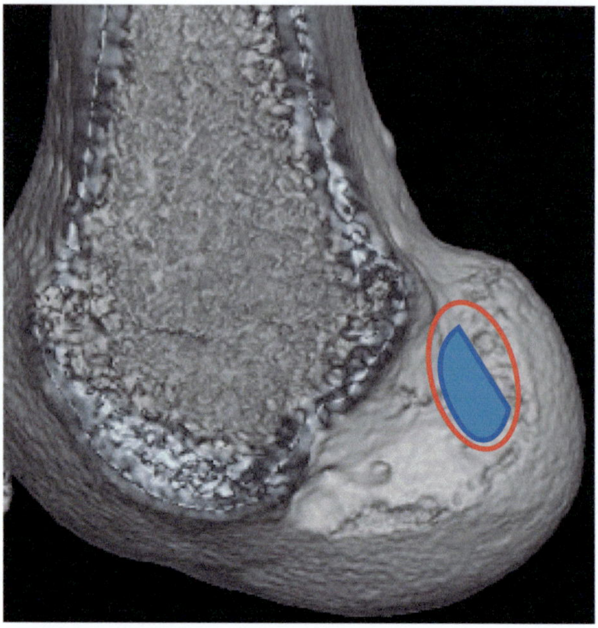

Fig. 7.9 3-D CT image of the lateral wall of the intercondylar notch showing a thick single bundle soft tissue graft inside a cylindrical femoral tunnel (right knee)
In a single bundle reconstruction using a multi-stranded hamstring tendon graft, even if a round tunnel (red oval) of 8–10 mm is created in the center of the femoral attachment area, it is deviated anteriorly-distally inside the tunnel (blue area) after tensioning the graft, resulting in a vertical graft

POINTS

1. Intraoperative radiography with guide pins inserted in the center of the attachment areas helps to reduce errors (Fig. 7.10).
2. Since the attachment areas are not round or elliptical, drilling with an 8–10 mm drill-bit makes protrusion of the tunnel apertures from the attachment area. Thus multiple smaller or rectangular tunnels are recommended.
3. The femoral condyles slip posteriorly while weight-bearing at the time of activities, leading to the repetitive loadings of the graft. Thus the anterior wall of the femoral tunnel and the posterior wall of the tibial tunnel are always stressed and eroded, since ossification of the graft inside the tunnels may not occur even in several years postoperatively. This makes the graft vertically over time, and may result in recurrence of instability.

Therefore, the posterior margin of the femoral tunnel should be aligned with the posterior superior margin of the femoral attachment area, and the anterior margin of the tibial tunnel aperture with the antero-medial margin of the tibial attachment area. In other words, anterior placement of the femoral tunnel or posterior placement of the tibial tunnel should be avoided

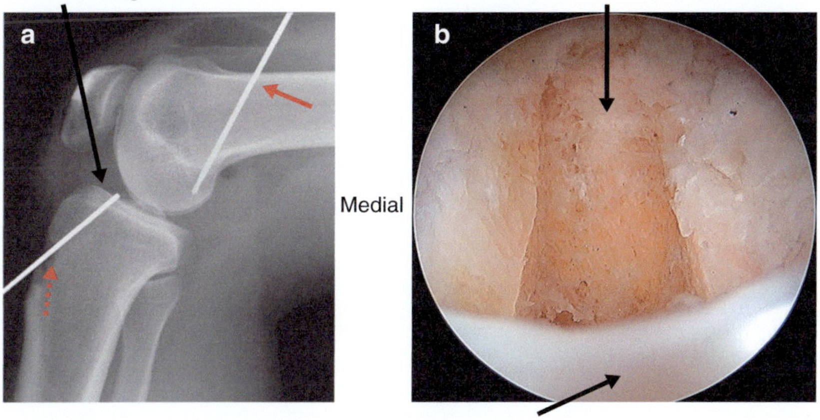

Fig. 7.10 Direction and aperture of the tibial tunnel
(**a**) Intraoperative lateral radiograph. A guide pin inserted in the center of the tibial attachment area (dashed arrow) is used as a reference to create a more accurate tibial tunnel. Also note the tip of the pin inserted into the femur is located at the center of the femoral attachment (arrow)
(**b**) Arthroscopic view of the rectangular tibial tunnel aperture and the transverse ligament (left knee). The transverse ligament (ligament connecting the anterior horn of the medial meniscus to the anterior segment of the lateral meniscus) sometimes makes it difficult to visualize the anterior ridge/Parsons' knob or the anterior border of the tibial tunnel aperture. In such a case with a well-developed transverse ligament, it is difficult to arthroscopically view the anterior margin of the tibial tunnel. Thus intraoperative lateral radiograph is helpful

7.4 Graft Fixation Under Tension

The final fixation of the graft under tension should be performed around 20° of knee flexion. This is because the tensioned posterolateral structures relatively preserve the femur-tibia positional relationship, although some posterolateral displacement of the tibia is inevitable when the graft is tensioned [11].

While it is generally accepted that some tension should be applied to the graft, there is no established theory on the amount of tension to be applied to the graft at the time of its fixation. Laxity match pretension (LMP: the amount of tension to be applied to the graft in order to restore the normal stability) could be a good reference. However, it varies depending on the surgical procedure and the graft. For correctly/anatomically-placed grafts, initial tension of 10–20 N (from a tensioner installed to the tibia/calf, not from the one held by the surgeon's hand) is considered appropriate [12] (Fig. 7.11).

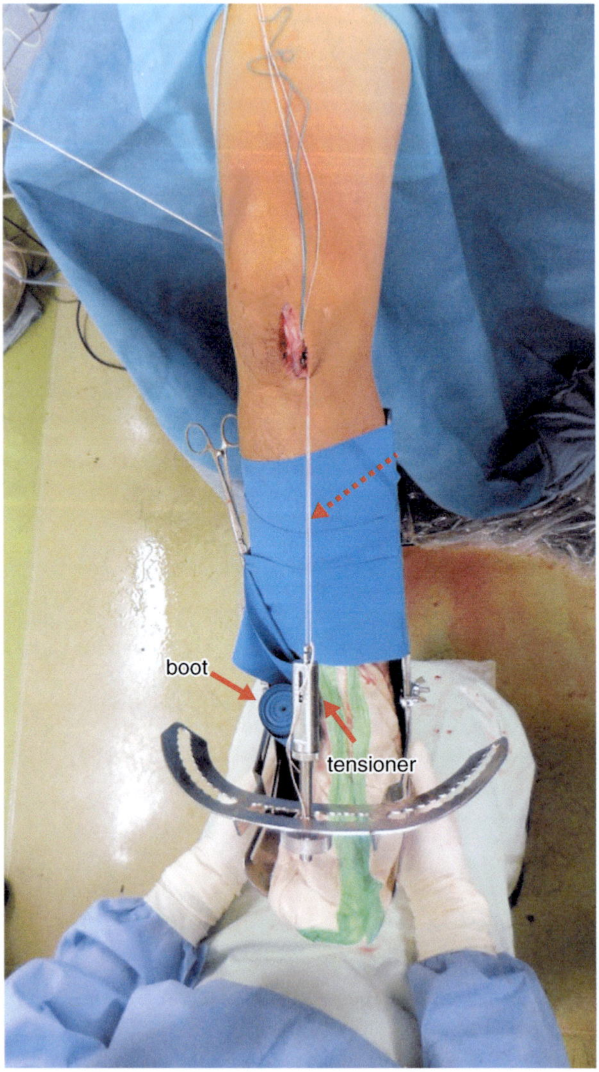

Fig. 7.11 *In situ* load relaxation of a graft using a boot using a tensioner with a metal tensioning boot bandage-fixed to the calf
After the graft is fixed on the femoral side, the sutures from the distal end of the graft (dashed arrow) is fixed to a tensioner installed to the boot. While monitoring the tension, the sutures are strongly pulled repeatedly for 10 times, leading to load relaxation of the construct. Once the tensioner reading has been stabilized at the desired value, the suture is fixed under the intended tension with DSP and a screw (Fig. 7.12)

While pull-out sutures over a button or around a screw post is popular for final graft fixation at the tibia, it is unknown how much tension is applied with this kind of fixation techniques. Thus the author developed the following two types of fixation devices/techniques to adjust the initial tension to the graft at the time of final fixation at the tibia: (1) double spike plate (DSP) + screw (Fig. 7.12) [13] for pull-out suture fixation; (2) bone-plug tensioning & fixation (BTF) system consisting of spike button (SB) + intra-bone plug screw (IBS) (Fig. 7.13) for pull-out screw fixation. Both are distributed by Smith-Nephew.

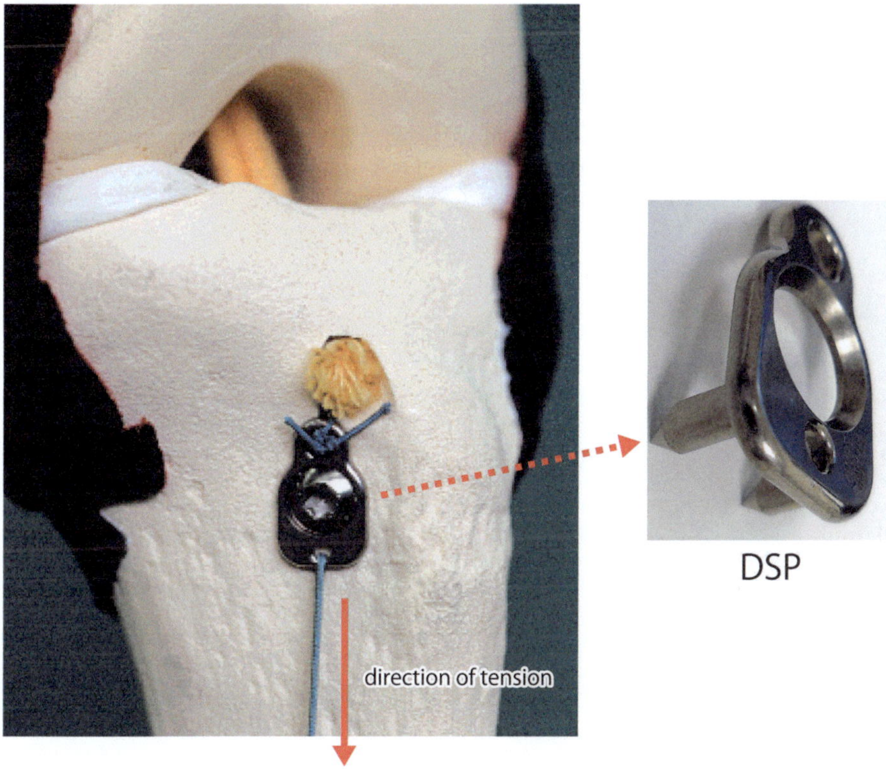

Fig. 7.12 Final graft fixation to the tibia under the intended residual tension with DSP and a screw (Figs. 8.1b and 8.6b)
The DSP is a small titanium plate with one/two top holes and one bottom hole and with two spikes on the underside of the bottom
At the time of using DSP for final fixation at the tibia, the graft sutures (the sutures from the distal end of the graft) are tied to the top hole(s) of DSP, and the tensioning sutures (the other sutures connected to the bottom hole of DSP) are tied to the tensioner. Tension is applied distally with the tensioning sutures. After load relaxation is achieved by repetitive manual strong pulls for 10 times, the bottom spikes are hammered into the tibia for temporary fixation. A screw is then inserted into the central hole for finalizing fixation. The tension is maintained throughout the fixation
(Adapted from Ref. [13])

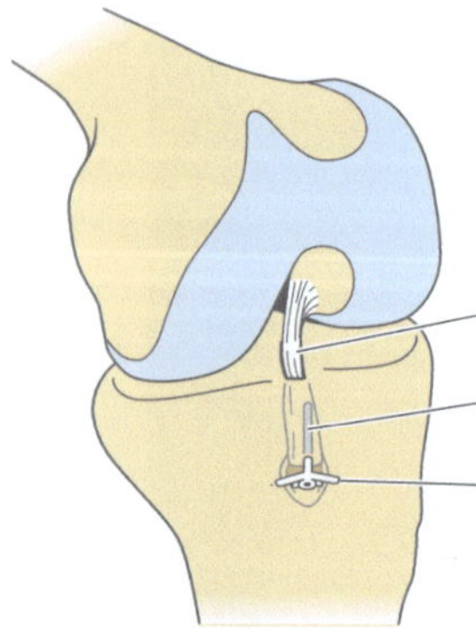

Fig. 7.13 Bone-plug tensioning & fixation (BTF) system for a bone-patellar tendon-bone (BTB) graft to the tibia under the intended residual tension (Figs. 11.5 and 11.11b)
BTF system consists of two components: Spike Button (SB) and Intra-Bone plug Screw (IBS). SB is a small rhombic plate with a central screw hole and two spikes on the undersurface. It is installed round the aperture of the tibial tunnel to serve as a rigid base for IBS pull-out fixation of the bone plug. IBS is inserted through the SB hole into the center of the bone plug inside the tibial tunnel. First, SB is driven into the tibial cortex around the tunnel aperture. Then, *in situ* pretensioning of a graft to achieve load relaxation is performed as described Figs. 7.11 and 12, and the tension is adjusted as intended. By advancing IBS into the bone plug in the tunnel through SB while the tension is monitored, the bone plug is pulled distally to tension the graft, leading to the final fixation under tension. This system cannot be used in case of a longer tendinous portion of the BTB graft, as the base of the distal bone plug is protruded out of the tibial tunnel

7.5 Remodeling Process of the Graft

1. **Knowledge from the animal experiments**

Transplanted tendon grafts, even viable autologous tissues, undergo the following processes: (1) ischemic necrosis, (2) remodeling as a new ligament by cellular and vascular invasion and (3) maturation [14)].

During the first few weeks after transplantation, the mechanical strength of the transplanted tendon graft is drastically reduced to 16–24% of that of the control ligament, due to fragmentation, resorption, neogenesis, and reorganization of collagen fibers. Subsequently, the mechanical strength of the transplanted tendon gradually increases until about 1 year after transplantation as it is modeled as a newly-formed ligament. When the graft reaches the maturation, the arrangement of collagen fibers is reorganized to the similar degree as normal. However, the mechanical strength (structural properties) does not return to normal and is assumed to recover only up to 70–80% (Fig. 7.14) [15].

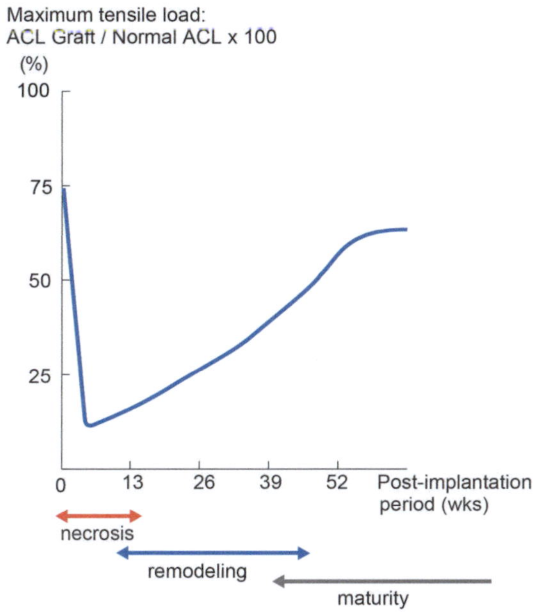

Fig. 7.14 Estimated mechanical strength of the reconstructed ACL with an autologous patellar tendon graft over time
The reconstructed ligament undergoes (1) ischemic necrosis, (2) remodeling and (3) maturation

The above-mentioned remodeling process of the reconstructed ACL was mainly inferred from the animal experiments. The data obtained from the animal model are not easy to interpret because the reconstructed ligaments are not protected by controlled postoperative rehabilitation. The transplanted grafts are stretched out in the early stage, and should be regarded as "failure model".

POINT

1. A viable autologous tissue transplanted into a joint as a reconstructed ACL undergoes ischemic necrosis along with drastic reduction of its strength. This means that the transplanted autologous tissue is merely a scaffold for inducing new tissue.
2. The tunnel apertures must be remained inside the anatomical attachment areas for correct graft placement. If they violate the border of the areas, normal structures are at risk to be damaged.

3. Allogeneic tendons can be used as grafts like autologous tendons as their antigenicity is little if treated properly. The author first performed several hundred cases of ACL reconstruction using fresh-frozen allogeneic tendon grafts with no complications. However, the procedure was discontinued because of shortage of graft supply. Currently, in some parts of Europe and the United States, cadaveric tissues including tendons are commercially avilable to be used routinely.

2. **Graft healing and remodeling in humans**

   The limited evidences are summarized below.

(1) The transplanted tendon as a reconstructed ACL increases in its cross-sectional area postoperatively, peaks and then decreases, reaching a plateau after 2–3 years (Chap. 9, Fig. 9.1). Its cross-sectional area remains about 1.2–1.8 times greater than that of the graft before transplantation [16, 17].
(2) Postoperative graft hypertrophy is more pronounced in the patellar tendon than in the hamstring tendon graft [17].
(3) The ultra-microstructure/collagen fibril profile of the tendon before transplantation shows a bimodal distribution (large diameter 90–140 nm and small diameter 30–80 nm) (Fig. 7.15a), whereas after transplantation, it shows a unimodal distribution consisting only of small diameter fibrils of 30–80 nm (Fig. 7.15b) [18] . This suggests that the transplanted tendon is transformed into a weakened tissue of different microstructure by the remodeling process, and that its mechanical properties are degraded after transplantation.
(4) An ACL graft reconstructed with the patellar tendon is not entirely remodeled even at 18 months after surgery, showing necrosis in part [19].
(5) An ACL graft reconstructed with the patellar tendon exhibited 87% of the maximum failure load of the healthy side at 8 months postoperatively [20].

POINT
The transplanted tendon/reconstructed ACL is replaced by newly-formed tissue and will not maintain its initial strength. The maximum failure load of the reconstructed ligament could be assumed to be 80–90% of that of the normal ACL.

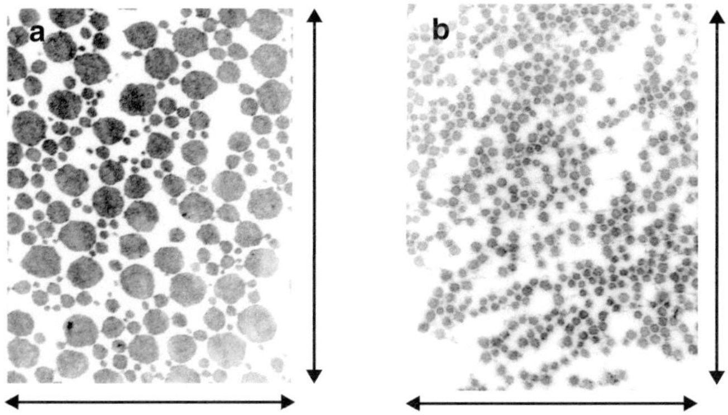

Fig. 7.15 Ultra-microstructure of the tendon and the reconstructed ligament (transmission electron microscope image) x 17360
(**a**) The collagen fibril profile of the tendon before transplantation shows a bimodal distribution. (large diameter: 90~140 nm and small diameter: 30–80 nm)
(**b**) The collagen fibril profile at 1 year-old ACL graft reconstructed with allogeneic tendon shows a unimodal distribution consisting only of small diameter fibers of 30–80 nm. This suggests a decrease in mechanical properties after transplantation

References

1. Kondo M. Experimental study on the formation of the anterior cruciate transfiguration of the knee joint. J Jpn Orthopaedic Assoc. 1979;53:521–33. (in Japanese)
2. Noyes FR, et al. Biomechanical analysis of human ligament grafts used in knee-ligament repairs and reconstructions. J Bone Joint Surg Am. 1984;66:344–52.
3. Stäubli HU, et al. Mechanical tensile properties of the quadriceps tendon and patellar ligament reconstructions. Am J Sports Med. 1999;27:27–34.
4. Shino K, et al. Replacement of the anterior cruciate ligament by an allogeneic tendon graft. An experimental study in the dog. J Bone Joint Surg Br. 1984;66:672–81.
5. Shino K, et al. Reconstruction of the anterior cruciate ligament using allogeneic tendon: long-term followup. Am J Sports Med. 1990;18:457–65.
6. Shino K, et al. Anatomic ACL reconstruction: rectangular tunnel/bone-patellar tendon-bone or triple-bundle/semitendinosus tendon grafting. J Orthop Sci. 2015;20:457–68.
7. Shino K, et al. The resident's ridge as an arthroscopic landmark for anatomical femoral tunnel drilling in ACL reconstruction. Knee Surg Sports Traumatol Arthrosc. 2010;18:1164–8.
8. Hiramatsu K, Shino K, et al. Contact area between femoral tunnel and interference screw in anatomic rectangular tunnel ACL reconstruction: a comparison of outside-in and trans-portal inside-out techniques. Knee Surg Sports Traumatol Arthrosc. 2018;26:519–25.
9. LaPrade CM, et al. Consequences of tibial tunnel reaming on the meniscus roots during cruciate ligament reconstruction in a cadaveric model, Part 1: the anterior cruciate ligament. Am J Sports Med. 2015;43:200–6.

10. Take Y, Shino K, et al. Early structural results after anatomic triple bundle anterior cruciate ligament reconstruction validated by tunnel location, graft orientation, and static anteroposterior tibia-femur relationship. Arthroscopy. 2018;34:2656–65.
11. Mae T, Shino K, et al. Optimization of graft fixation at the time of anterior cruciate ligament reconstruction. Part II: effect of knee flexion angle. Am J Sports Med. 2008;36:1094–100.
12. Mae T, Shino K, et al. Anatomic double-bundle anterior cruciate ligament reconstruction using hamstring tendons with minimally required initial tension. Arthroscopy. 2010;26:1289–95.
13. Shino K, et al. Graft fixation with pre-determined tension using a new device, the double spike plate. Arthroscopy. 2002;18:908–11.
14. Clancy WG, et al. Anterior and posterior cruciate ligament reconstruction in rhesus monkeys. J Bone Joint Surg Am. 1981;63:1270–84.
15. Butler DL, et al. Mechanical properties of primate vascularized vs. nonvascularized patellar tendon grafts:changes over time. J Orthop Res. 1989;7:68–79.
16. Kinugasa K, Shino K, et al. Cross-sectional area of hamstring tendon autograft after anatomic triple-bundle ACL reconstruction. Knee Surg Sports Traumatol Arthrosc. 2017;25:1219–26.
17. Kinugasa K, Shino K, et al. Chronological changes in the cross-sectional area of the bone-patellar tendon-bone autograft after anatomic rectangular tunnel ACL reconstruction. Knee Surg Sports Traumatol Arthrosc. 2021;29:3782–92.
18. Shino K, et al. Collagen fibril populations in human anterior cruciate ligament allografts. Electron microscopic analysis. Am Sports Med. 1995;23:203–9.
19. Delay BS, et al. Observations on a retrieved patellar tendon autograft used to reconstruct the anterior cruciate ligament. A case report. J Bone Joint Surg Am. 2002;84:1433–8.
20. Beynnon BD, et al. Evaluation of knee joint laxity and the structural properties of the anterior cruciate ligament graft in the human: a case report. Am J Sports Med. 1977;25:203–6.

Chapter 8
Anatomical ACL Reconstruction

This book is not for operative procedure, so only a brief description on the anatomical ACL reconstructions is given. The author's own surgeries are available on the Internet and can be accessed at any time at the homepage of Yukioka Hospital (Sports Orthopedic Center) (http://www.yukioka.or.jp/medical/treatment/shino.html) or at VuMedi (https://www.vumedi.com/search/?q=Shino).

8.1 Basic Concept

The normal ACL is located in the intercondylar notch and is attached to the two anatomical attachment areas: one, the posterior-superior portion of the lateral femoral condyle; the other, the anteromedial part of the tibial plateau. It does not impinge on the intercondylar notch or the posterior cruciate ligament during the entire range of motion except for hyperextension. An anatomical ACL graft is realized by the anatomical ACL reconstruction procedure in which the two attachment areas are connected with a robust graft. It should show no impingement to the other structures within normal range of motion, and become taut in extension.

8.2 Anatomical Reconstruction Techniques

1. Anatomical Triple Bundle ACL Reconstruction (ATB/ ACLR) with Autologous Hamstring Tendon [1]

The reconstruction mimics the normal ligament composed of three fiber bundles as closely as possible to anatomically resemble the normal ligament (Fig. 8.1).

K. Shino, *Essence of Anterior Cruciate Ligament*,
https://doi.org/10.1007/978-981-99-6536-6_8

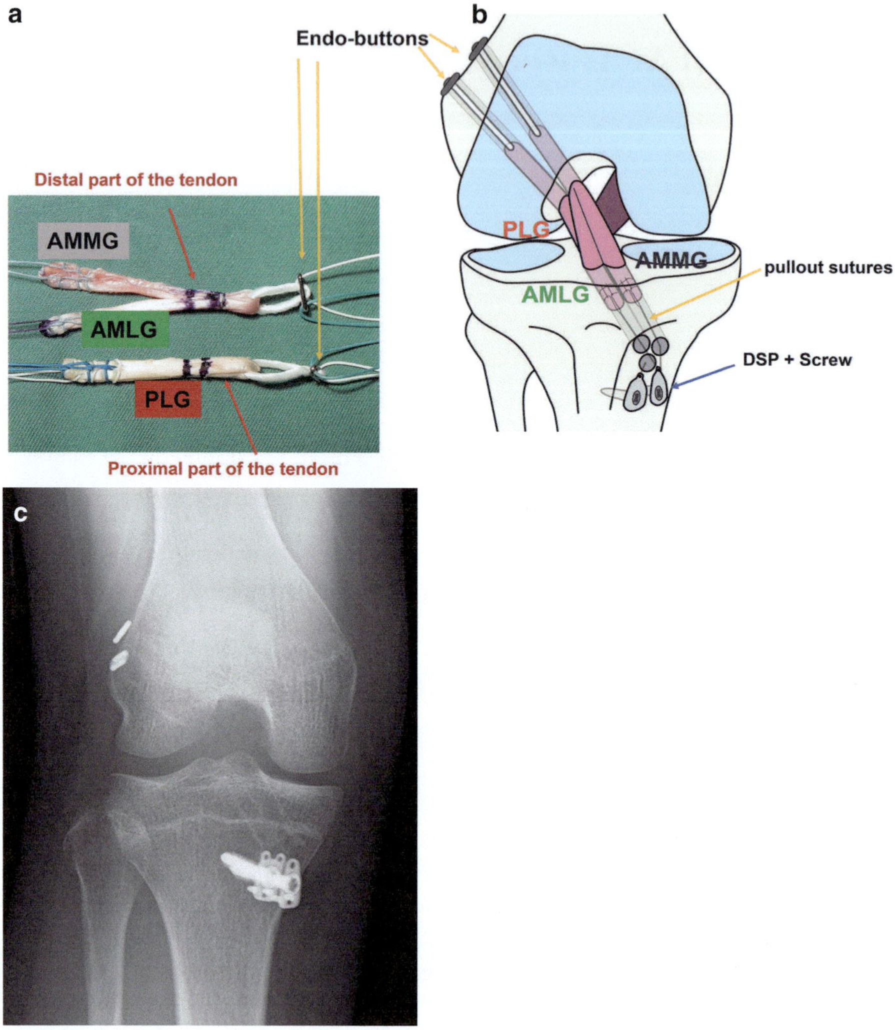

Fig. 8.1 Anatomical triple bundle reconstruction using autologous hamstring tendon (semitendinosus tendon)
DSP, double spike plate; *AMMG*, Anteromedial Medial Graft; *AMLG*, Anteromedial Lateral Graft; *PLG*, Posterolateral Graft
(**a**) Prepared tendon graft
(**b**) Schema of the procedure
(**c**) Postoperative X-p

The whole semitendinosus tendon is used as a principal graft. The total tendon is harvested, cut transversely. Each tendon is folded into two double-looped tendon grafts of 5–6 mm in diameter and 6–7 cm in length (Fig. 8.1a). In cases of hamstring tendon hypoplasia, one folded semitendinosus tendon graft and one folded gracilis tendon graft are used. The cross-sectional area (transverse diameter) of these tendons should be evaluated by MRI before surgery [2] (Fig. 8.2).

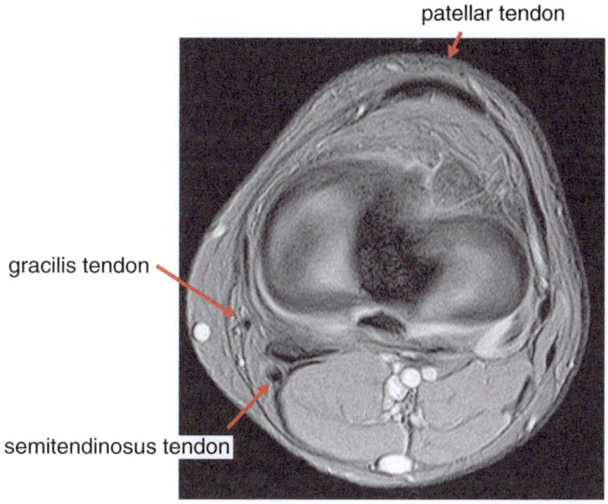

Fig. 8.2 Preoperative evaluation of tendon grafts by MRI (T2* -weighted, axial image at tibial plateau level)
In this case, the maximum diameter of the semitendinosus tendon was 5 mm, making it possible to prepare the triple bundle graft consisting of only the semitendinosus tendon. When the maximum diameter is less than 4 mm, additional graft harvest may be considered

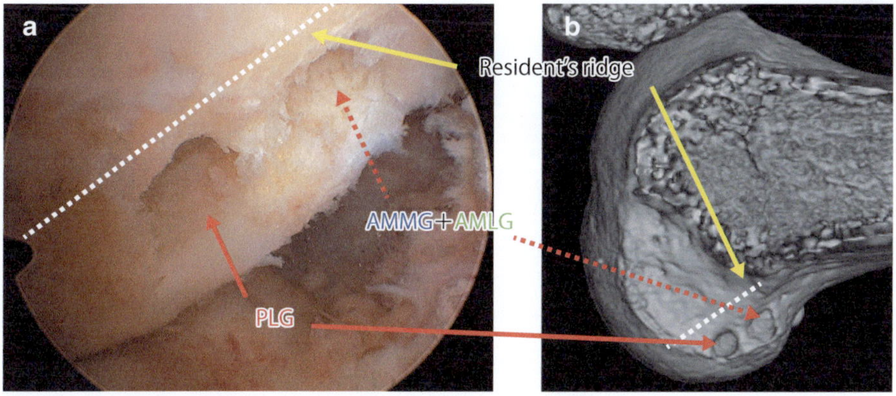

Fig. 8.3 Two small tunnels inside the ACL femoral attachment area of the right knee
(**a**) Arthroscopic view
(**b**) 3-D CT image
AMMG and AMLG are placed in the superior/proximal tunnel and PLG is introduced in the inferior/distal tunnel

Two 5- to 6-mm diameter tunnels were created at the ACL attachment area of the femur: one superior/proximal tunnel for the two-limb graft of an anteromedial-medial graft(AMMG) and an anteromedial-lateral graft (AMLG); the other inferior/distal one for the single bundle posterolateral graft (PLG) consisting of one double-looped tendon. (Fig. 8.3).

Three tunnels are created at the tibial attachment site (Fig. 8.4). An AMMG and AMLG are placed in the anterior 4.5–5 mm diameter tunnel for each, and a PLG is placed in the posterior 5–6 mm diameter tunnel. This mimics each of the three fiber bundles of the normal ACL.

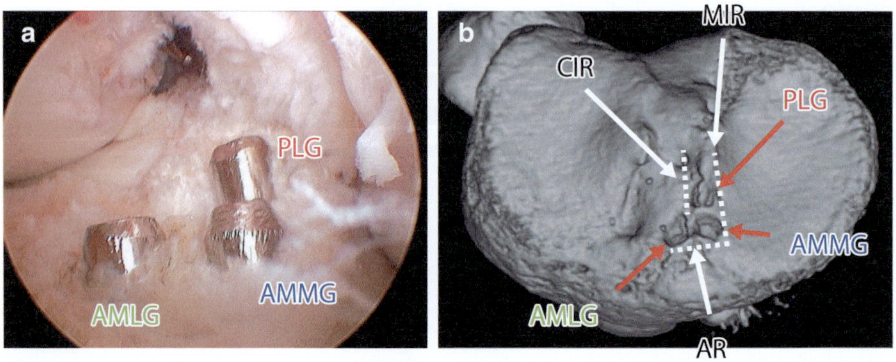

Fig. 8.4 Three small tunnels inside the ACL tibial attachment area of the right knee
(**a**) Arthroscopic view
(**b**) 3-D CT image
MIR Medial Intercondylar Ridge, *CIR* Central Intercondylar Ridge, *AR* Anterior Ridge/Parsons' knob

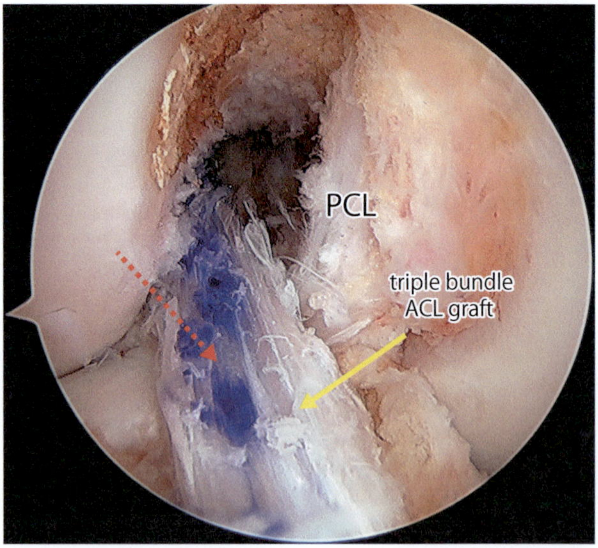

Fig. 8.5 Anatomical triple bundle ACL graft with semitendinosus tendon (right knee)
The AMLG is stained with blue dye (dashed arrow) for identification.
PCL, Posterior cruciate ligament

As a rule, the more robust distal portion of the semitendinosus tendon is assigned to AMMG and AMLG, while the proximal fascia-like portion is used for PLG.

The reconstructed ligament obtained by this procedure is morphologically very close to the normal ACL [3] (Fig. 8.5).

The problem with the reconstructions using soft tissue graft such as the present procedure is that solid graft-femoral tunnel integration is unlikely to be completed [4].

2. Anatomical Rectangular Tunnel ACL Reconstruction with autologous bone-patellar tendon-bone graft [5] (ART ACLR with BTB Graft) (Fig. 8.6).

A patellar tendon of 8–13 mm width with attached bone plugs of 15 mm length is used as a graft. A parallelepiped bony tunnel with a rectangular cross section of 5–6.5 × 10–13 mm is created at the ACL attachment areas of the tibia and the femur to anatomically mimic the normal ligament (Figs. 8.7 and 8.8). Although this technique

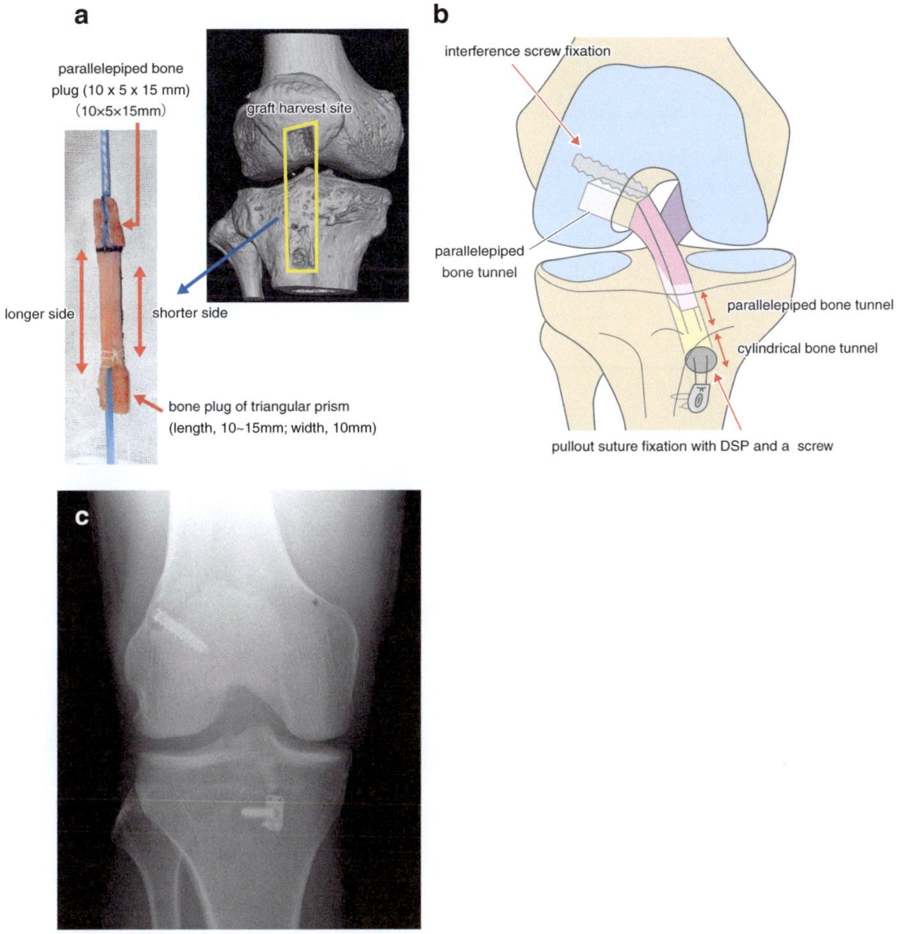

Fig. 8.6 Anatomical rectangular tunnel ACL reconstruction with central portion of the medial 1/2 of the patellar tendon
The graft is introduced into the tunnels with a rectangular cross section at the ACL attachment areas in the tibia and femur to mimic the normal ACL
DSP Double Spike Plate (See Fig. 7.12)
(**a**) Prepared bone-patellar tendon-bone (BTB) graft
(**b**) Schema of the procedure
(**c**) Postoperative X-p

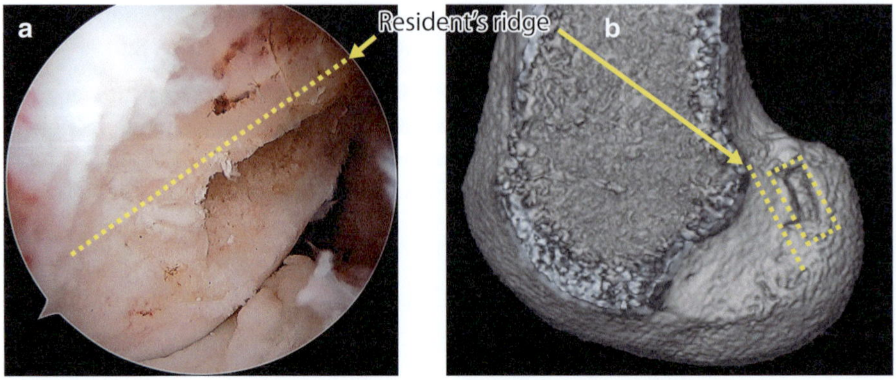

Fig. 8.7 Rectangular/parallelepiped tunnel created at the right knee femoral attachment site
(**a**) Arthroscopic view
(**b**) 3-D CT image
The tunnel aperture is located inside the attachment area located posterior-proximal margin on the lateral wall of the intercondylar notch and posterior to the resident's ridge

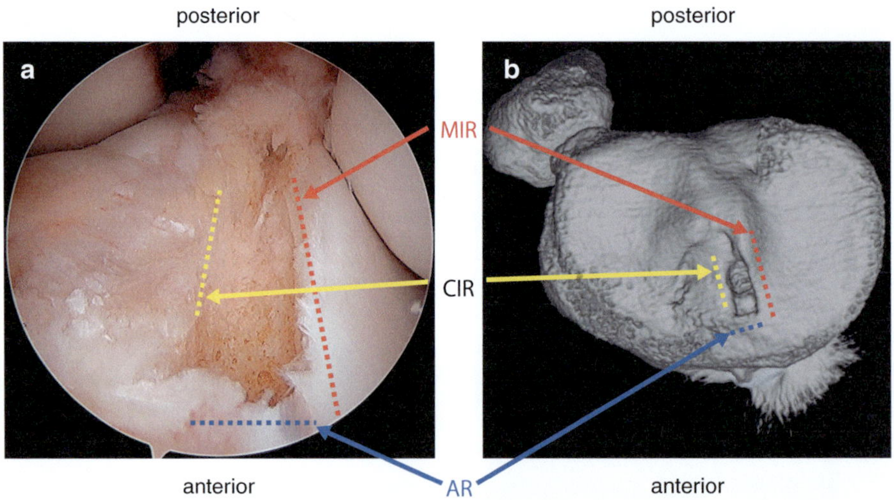

Fig. 8.8 Rectangular/ /parallelepiped tunnel aperture is created inside the tibial attachment area of the right knee
(**a**) Arthroscopic view
(**b**) 3-D CT image
The bony landmarks [medial intercondylar ridge (MIR), anterior ridge (AR), and central intercondylar ridge (CIR)] on the tibial plateau are confirmed arthroscopically

uses a single bundle graft, it can mimic the internal fiber arrangement of the normal ligament to a great extent, resulting in the internal fiber arrangement similar to the anatomic triple-bundle/double-bundle reconstruction technique described above (Fig. 8.9). In addition, there is almost no undesirable gap between the tendinous portion of the graft and the tunnel wall, and early graft-tunnel integration can be expected [6].

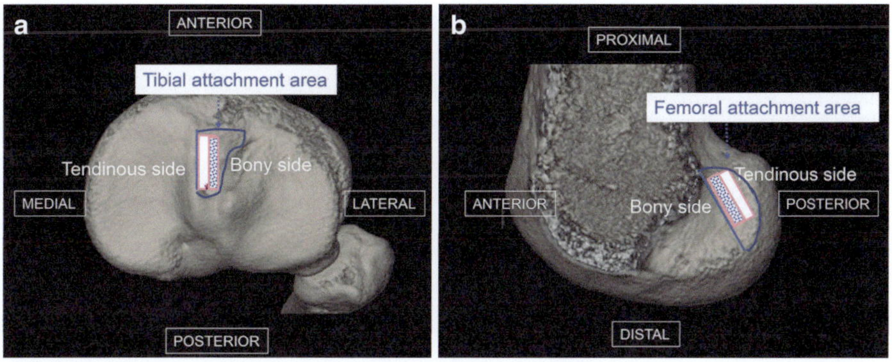

Fig. 8.9 Deployment of the bone plugs of a BTB graft in the anatomical rectangular tunnel reconstruction. (**a**) 3-D CT image of the tibial plateau; (**b**) 3-D CT image of the lateral wall of the intercondylar notch. Note this deployment of each bone plug to mimic the internal fiber arrangement of the normal ACL with the patellar tendon (right knee)

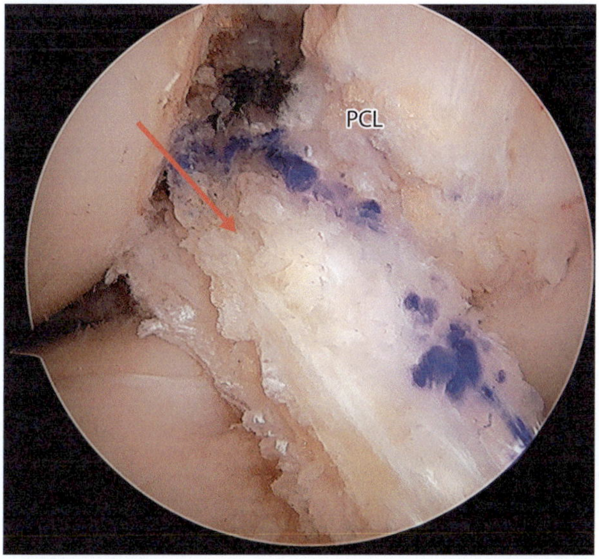

Fig. 8.10 Arthroscopic view of the anatomically-placed patellar tendon graft (right knee)
The graft does not impinge on the other structures including PCL
PCL, Posterior cruciate ligament

Within the femoral tunnel, the cancellous side of the bone plug (taken from the tibial tubercle) should be faced to the anterior wall of the tunnel, and the tendon-bone junction is aligned with the aperture of the tunnel. The patellar tendon is placed so that its ventral aspect is medial. The other bone plug (taken from the patella) is placed in the tibial tunnel with the cancellous side faced laterally (Figs. 8.9 and 8.10).

3. Anatomic rectangular tunnel reconstruction using autologous quadriceps tendon with a bone plug on one end

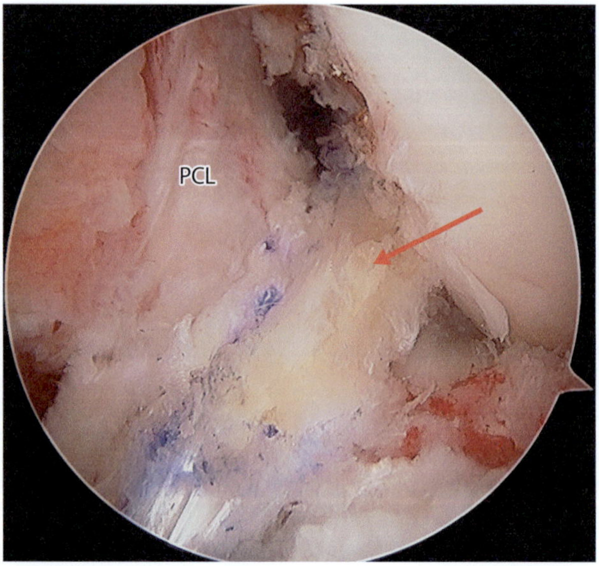

Fig. 8.11 Arthroscopic view of the anatomically-placed quadriceps femoris tendon graft (left knee) The graft does not impinge on the other structures including PCL. *PCL* Posterior cruciate ligament

A 6 cm or longer quadriceps tendon of 8–13 mm width, along with a 15-mm long bone plug from the patella on one end, is harvested.

This procedure is similar to the ART ACLR with BTB Graft, but the distal portion of the graft inside the tibial tunnel is a free tendon.

Within the femoral tunnel, the cancellous side of the bone plug (taken from the patella) should be faced to the anterior wall of the tunnel, and the tendon-bone junction is aligned with the aperture of the tunnel. The quadriceps tendon is placed so that its ventral aspect is medial (Fig. 8.11).

8.3 Biomechanics of Anatomical Reconstruction Procedures

1. Robotic knee simulator for biomechanical testing

The robotic simulator developed by Fujie et al. (Fig. 8.12) can accurately reproduce the trajectory of 6-DOF joint motion [7]. This makes it possible to precisely manipulate the knee to strictly perform the laxity examinations including the Lachman test.

It is equipped with a universal force/moment sensor (UF/MS). With the principle of superposition, it is possible to evaluate force/moment exerted on each structure of the knee joint.

2. Comparison of stabilizing efficiency among the surgical procedures

Since ligament reconstruction is performed to restore joint stability, the transplanted tendon/reconstructed ligament is finally fixed under a certain amount of tension. The tension applied to the graft at the time of final fixation is called the

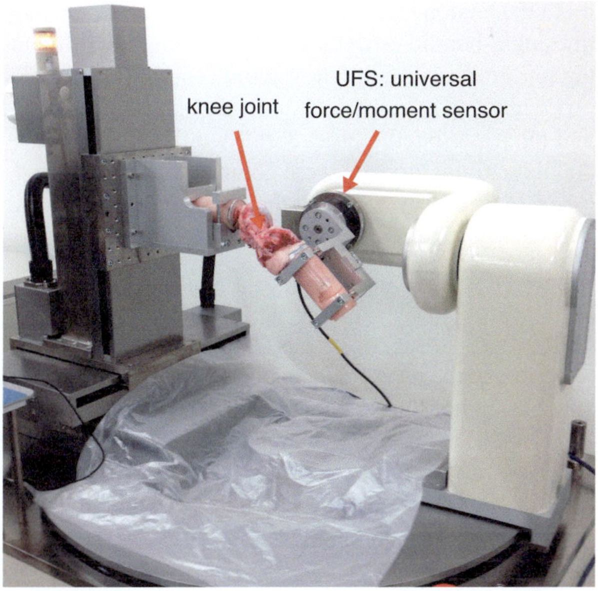

Fig. 8.12 A robotic knee simulator for biomechanical testing

initial (fixation) tension. Laxity match pretension (LMP) is defined as the initial tension required to restore the stability equivalent to the normal knee. Concretely, this value is determined as the tension required to achieve the same amount of anterior tibial displacement during the Lachman test (100 N anterior tibial drawer force at 30° flexion) in the normal knee [8]. This value varies depending on the procedure as well as the graft, and can be used as an indicator of stabilizing efficiency among the surgical procedures. When the normal ACL is used as the graft in a strictly anatomical reconstruction, the required initial tension is 0 N. Thus, the smaller the LMP, the higher the stabilizing efficiency of the reconstruction.

(1) Effect of tunnel location in the reconstructive procedures using a patellar tendon graft

In the cadaveric knee experiments, a comparison of stabilizing efficiency between procedures with a 10-mm wide autologous patellar tendon graft was performed using LMP as an indicator. For the anatomical rectangular tunnel (ART) reconstruction, the femoral tunnel was created in the anatomical attachment area located posterior-proximal to the resident's ridge. For the isometric round tunnel (IRT) reconstruction, the femoral tunnel was created anterior-proximal to the anatomical attachment area [9]. The tibial tunnel location was the same between the procedures. The mean LMP was 8.6 N for the former procedure, while the mean LMP for the latter was 34.8 N. The stabilizing efficiency of the ART procedure was clearly superior to that of the IRT.

(2) Comparison of single-, double-, and triple-bundle procedures in the anatomical reconstruction using hamstring tendon graft [10]

In the anatomical round tunnel reconstruction using the hamstring tendon grafts, the single-bundle procedure (a single-bundle graft consisting of two double-looped tendon grafts via a single femoral and a single tibial tunnel), the double-bundle procedure (two double-looped tendon grafts via two femoral and two tibial tunnels) and the triple-bundle procedure (two double-looped tendon grafts via two femoral and three tibial tunnels) were compared. The average LMP was 26.3 N, 11.2 N, and 6.8 N, respectively, showing the stabilizing efficiency of the triple-bundle procedure was clearly superior.

POINT

1. Since the anatomical attachment areas are not round, the single rectangular tunnel or multiple small tunnel techniques are desirable to keep the tunnel apertures within the attachment areas. This makes it possible for the graft to closely mimic the normal ACL without its impingement to the other structures. In addition, this leads to the increased graft-tunnel contact area to enhance biological healing of the graft.
2. As the anatomically-placed grafts show higher stabilizing efficiency, the procedure can be performed with smaller initial tension to the grafts.
3. In non-anatomical reconstructions, higher initial tension is required to restore the stability at the time of graft fixation, resulting in abnormal femur-tibia positional relationship (posterolateral displacement of the tibia, valgus alignment) [9]. Those that require excessive initial tension not only overload the grafts but also lead to articular cartilage degeneration.

8.4 An Acceptable Non-anatomical Reconstruction Technique

A non-anatomical reconstruction may be indicated for patients in growing age with open epiphyses, as a femoral tunnel cannot be created to avoid growth disturbance. The iliotibial band is transected proximally to create a distally-pedicled graft. A tibial tunnel is created in the anatomical attachment area. On the femoral side, without creating a bony tunnel, the proximal end of the iliotibial band graft is introduced into the joint through the posterior joint capsule via over-the-top of the lateral condyle, and then pulled through the tibial tunnel to the anterior tibial cortex [11]. The reconstructed ligament is not strictly anatomical but fairly close to the normal ligament (Fig. 8.13) [12]. This over-the -top recopnstruction could be better performed with the two doubled looped semitendinosus tendons [13].

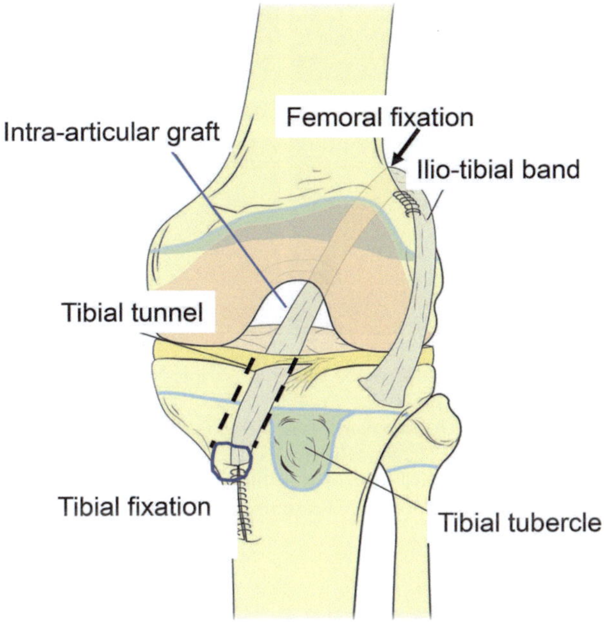

Fig. 8.13 A non-anatomical reconstruction with the iliotibial band via over-the-top of the lateral femoral condyle: an acceptable non-anatomical procedure
The iliotibial band is transected proximally to create a distally-based pedicled graft. A tibial tunnel is created in the anatomical attachment area. On the femoral side, without creating a bony tunnel, the proximal end of the iliotibial band graft is introduced into the joint through the posterior joint capsule via over-the-top of the lateral condyle, and then pulled through the tibial tunnel to the anterior tibial cortex
(Based on Ref. [12])

8.5 A Future Anatomical Reconstruction Technique: In-lay Method

As previously mentioned, the greatest weakness of the current ACL reconstruction techniques with tendon grafting via tunnels is the graft bending angle (GBA) at the intra-articular aperture of the femoral tunnel. This angle causes stress concentration at the anterior margin of the aperture, resulting in widening and migration of the tunnel (Figs. 8.14 and 8.15) [14]. In the case of tendon grafts with bone plugs, the bone-tendon junction can be aligned with the anterior margin of the bony tunnel to somewhat reduce the stress to the graft.

To overcome this problem, an in-lay technique that minimizes GBA by creating a shallow socket that matches the shape of the attachment may be considered [15] (Fig. 8.16). However, when using a 10-mm wide patellar tendon, the area of the socket bottom is as small as 85 mm [2]. This small contact area makes it difficult to achieve sufficient initial fixation strength of the graft to the socket.

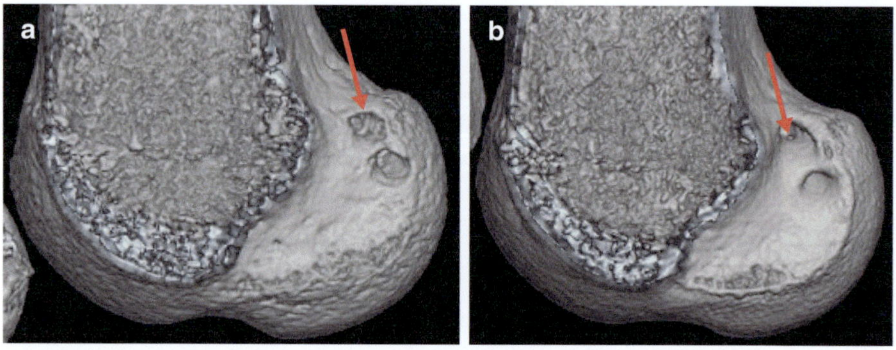

Fig. 8.14 Femoral tunnel apertures after the anatomical triple-bundle reconstruction using hamstring tendon graft (right knee, 3-D CT image)
(**a**) At 3 weeks postoperatively. The two tunnels are created within the attachment area
(**b**) At 17 months postoperatively. The tunnels have expanded anteriorly and distally. This enlargement/bony erosion is considered due to the stress by the graft which could not be transformed into bone. Revision ACL reconstruction for such a case requires some technical modifications such as use of thicker bone plugs

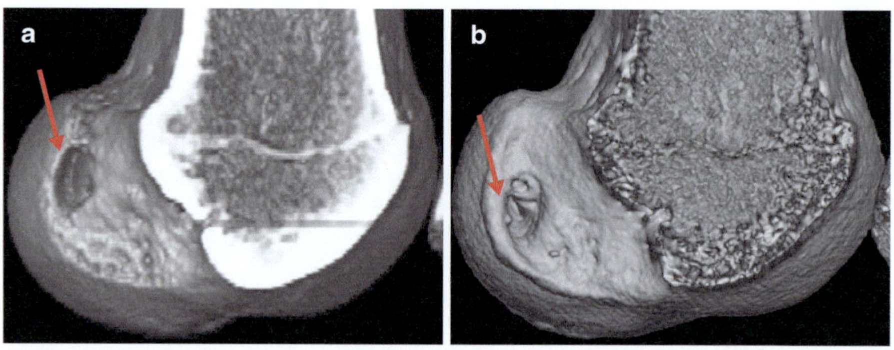

Fig. 8.15 Femoral tunnel aperture after anatomic rectangular tunnel ACL reconstruction using a bone-patellar tendon-bone graft (left knee, 3-D CT image)
(**a**) At one month after surgery
(**b**) At 8 years after surgery
The tunnel aperture shifted anteriorly and distally over years. Repetitive antero-distal pulling force to the graft during activities might have caused erosion of the antero-distal wall of the tunnel, as the bone is assumed weaker than the tendon. Since the bone plug was stabilized within the tunnel due to bony fusion, new bone formation may take place behind the shifted tunnel aperture

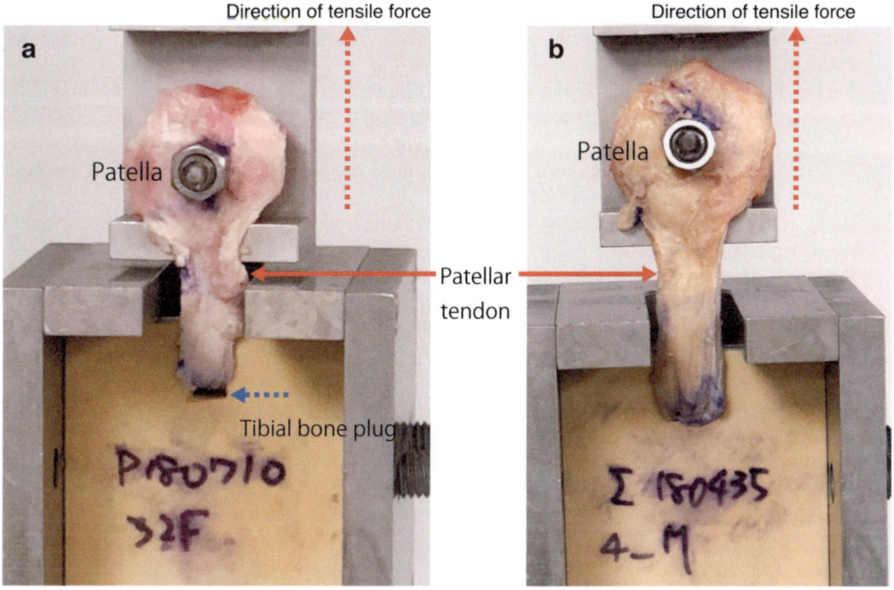

Fig. 8.16 Tensile tests to evaluate fixation strength of the bone plug of the bone-patellar tendon-bone specimen to the sawbone: Tunnel vs. In-lay technique
(**a**) Tunnel technique: A tibial bone plug is pulled into a 5 × 10 mm rectangular hole of full length and fixed with pullout sutures over a button. Despite snug fitting of the tunnel-bone plug, a void can be seen at the inferior part of the tunnel (blue dashed arrow)
(**b**) In-lay technique: The cancellous surface of the tibial bone plug is placed in the shallow socket of 5-mm depth and fixed with pullout sutures through small holes over a button. The bone plug appears to be attached to the bone like a normal ligament

References

1. Shino K, et al. Anatomic anterior cruciate ligament reconstruction using two double-looped hamstring tendon grafts via twin femoral and triple tibial tunnels. Oper Tech Orthop. 2005;15:130–4.
2. Hamada M, Shino K, et al. Cross-sectional area measurement of the semitendinosus tendon for anterior cruciate ligament reconstruction. Arthroscopy. 1998;14:696–701.
3. LaPrade CM, et al. Consequences of tibial tunnel reaming on the meniscul roots during cruciate ligament reconstruction in a cadaveric model, Part 1: the anterior cruciate ligament. Am J Sports Med. 2015;43:200–6.
4. Toritsuka Y, Shino K, et al. Arthroscopic evaluation of ACL grafts reconstructed with the anatomical two-bundle technique using hamstring tendon. Knee Surg Sports Traumatol Arthrosc. 2007;15:720–8.

5. Shino K, et al. Rectangular tunnel double-bundle anterior cruciate ligament reconstruction with bone-patellar tendon-bone graft to mimic natural fiber arrangement. Arthroscopy. 2008;24:1178–83.
6. Suzuki T, Shino K, et al. Early integration of a bone plug in the femoral tunnel in rectangular tunnel ACL reconstruction with a bone-patellar tendon-bone graft: a prospective computed tomography analysis. Knee Surg Sports Traumatol Arthrosc. 2011;19(Suppl 1):S29–35.
7. Fujie H, et al. The use of robotics technology to study human joint kinematics: a new methodology. J Biomech Eng. 1993;115:211–7.
8. Burks RT, et al. Determination of graft tension before fixation in anterior cruciate ligament reconstruction. Arthroscopy. 1988;4:260–6.
9. Suzuki T, Shino K, et al. Biomechanical comparison between the rectangular-tunnel and the round-tunnel anterior cruciate ligament reconstruction. Arthroscopy. 2014;30:1294–302.
10. Suzuki T, Shino K, et al. A biomechanical comparison of single-, double-, and triple-bundle anterior cruciate ligament reconstructions using a hamstring tendon graft. Arthroscopy. 2019;35:896–905.
11. Shino K, et al. How to handle a poorly placed femoral tunnel. In: Marx RG, editor. Revision ACL reconstruction: indications and technique. New York: Springer; 2014. p. 87–96.
12. Bertoia JT, et al. Anterior cruciate reconstruction using the MacIntosh lateral-substitution over-the-top repair. J Bone Joint Surg Am. 1985;67:1183–8.
13. Shiwaku K, Shino K, et al. A Biomechanical comparison of 2 over-the-top anterior cruciate ligament reconstruction techniques: A cadaveric study using a robotic simulator. Orthop J Sports Med. 2022;10:23259671221139876.
14. Tachibana Y, Shino K, et al. Femoral tunnel enlargement after anatomic anterior cruciate ligament reconstruction: bone-patellar tendon-bone/single rectangular tunnel versus hamstring tendon /double tunnels. J Orthop Sci. 2018;23:1011–8.
15. Wirth CJ, et al. Reconstruction of the anterior cruciate ligament. A new positioning and fixation technique. Am J Sports Med. 1990;18:154–9.

Chapter 9
Rehabilitation

9.1 Introduction

Rehabilitation after ACL reconstruction is important to normalize knee function as much as possible and return to sports by regaining range of motion and flexibility, and re-educating the atrophied muscles without compromising the reconstructed ligament. Even if an accurate anatomical reconstruction is performed, good results cannot be expected without appropriate rehabilitation. However, there is no magical rehabilitation that leads to a good outcome after non-anatomical reconstructions. The concrete rehabilitation described in this chapter is also applicable as conservative treatment of patients who cannot undergo surgery for some reason.

Programs should be implemented according to the remodeling/healing process of the reconstructed ligament. However, the evidences extrapolated from animal experiments to be used as a reference, are not simply applicable to humans, not only because there are species-differences between animals and humans, but because uncontrollable postoperative rehabilitation in animals consistently leads to stretching-out of the grafts (so-called "failure model").

While the remodeling/healing process of the transplanted grafts in humans is not entirely known, based on the results of animal experiments, it is generally thought that the transplanted graft undergoes the following phases: necrosis, degradation due to loss of cells; remodeling, metabolic hypertrophy with the invasion of blood vessels and cells; maturation, steady state/plateau without hypermetabolism (Fig. 7.14). In our study using MRI to investigate the temporal changes in the cross-sectional area of the transplanted grafts in humans, it took almost 3 years for the hypertrophied transplants to settle down and reach a plateau (Fig. 9.1), suggesting that it takes about 3 years for the grafts to biologically maturate [1, 2]. As a program of 3 years of postoperative rehabilitation followed by return to sports is unacceptable to everyone, as a realistic plan, we adopt a standard program of 9 months with

K. Shino, *Essence of Anterior Cruciate Ligament*,
https://doi.org/10.1007/978-981-99-6536-6_9

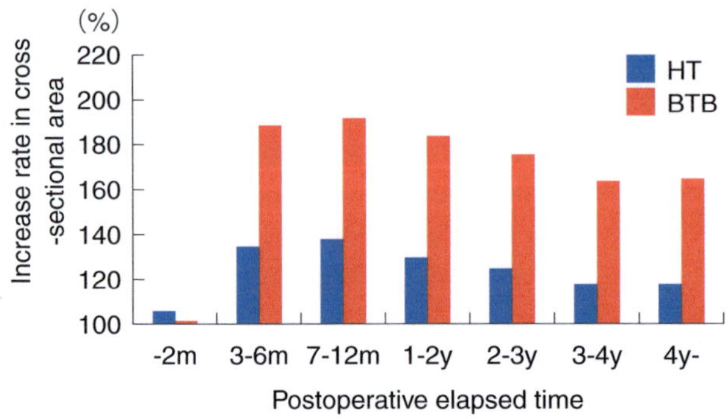

Fig. 9.1 Transition of transplanted tendon/reconstructed ligament cross-sectional area over time after ACL reconstruction in humans
It took almost 3 years for the hypertrophied graft to settle and reach a steady state/plateau. The BTB (patellar tendon) had a greater increase rate than the HT (hamstring tendon), but the time required to reach steady state was similar. (Adapted from Refs. [1, 2])

a minimum of 6 months before returning to sports, and categorized the patients into the following 4 phases:

- Phase I: recovery period, up to 6 weeks postoperatively
- Phase II: early training period, up to 3 months postoperatively
- Phase III: late training period (equivalent to the athletic rehabilitation phase), 3–6 months after surgery.
- Phase IV: return to sports period, after 6 months postoperatively

In addition, weight-bearing and range-of-motion exercises may be delayed due to additional treatments including concomitant meniscus repair.

9.2 Recovery Period [Up to 6 Weeks After Surgery (Fig. 9.2)]

Emphasis is on preventing arthrofibrosis. Restricted range of motion not only causes severe functional disability, but also induces osteoarthritis [3] . Therefore, the main goal during this period is to normalize the range of motion to prevent joint contractures/arthrofibrosis. Early range-of-motion training is essential for this purpose. However, the graft-femoral tunnel integration is obstructed under the presence of a space behind the graft inside the femoral tunnel (Fig. 7.8) if early range-of-motion exercise is permitted. Thus, a antinomy arises between early range-of-motion exercise to prevent arthrofibrosis and the immobilization necessary for graft- tunnel wall integration. As a practical compromise, immobilization in mild flexion (10–20°) with a soft brace (Fig. 9.3) is performed for a minimal period (1–2 weeks). Specifically, in a case who underwent the procedure using a bone-patellar

medical/athletic	medical rehabilitation					
elapsed postoperative period	0 weeks postoperatively	1 weeks postoperatively	2 weeks postoperatively	3 weeks postoperatively	4 weeks postoperatively	5 weeks postoperatively
rehabilitation category	recovery phase					
ROM(伸展/屈曲)	passive patellar mobilization → (continues to end) Soft brace immobilization →	BTB −10/75°	BTB, HT −5/90°	0/105°	0/120°	0/135°
Weight bearing/WB	full weight bearing →		1/3 WB	2/3 WB	Full WB	
muscle strengthening (OKC) Quad.	quad setting, SLR → (continues to end, dashed from 5 weeks)		leg extension (90° to 60°, isotonic) →			
muscle strengthening (OKC) Ham.		No restriction for BTB-grafted cases				
muscle strengthening (CKC)				bridge	half sitting exercise, one leg standing, calf raise	squatting on both legs
endurance training					underwater walking	stationary bicycle (no load) swimming (other than breaststroke)
posture control						
shock-absorbing control						
high power training						
running						
agility						
competition						

BTB: ACL reconstruction with bone-patellar tendon-bone graft; HT: reconstruction with hamstring tendon graft; OKC: open kinetic chain; CKC: closed kinetic chain; Quad.: quadricep femoris muscle; Ham.: hamstring muscle

Fig. 9.2 Phase I: Rehabilitation program in the recovery period (immediate postoperative period to 6 weeks postoperative period)

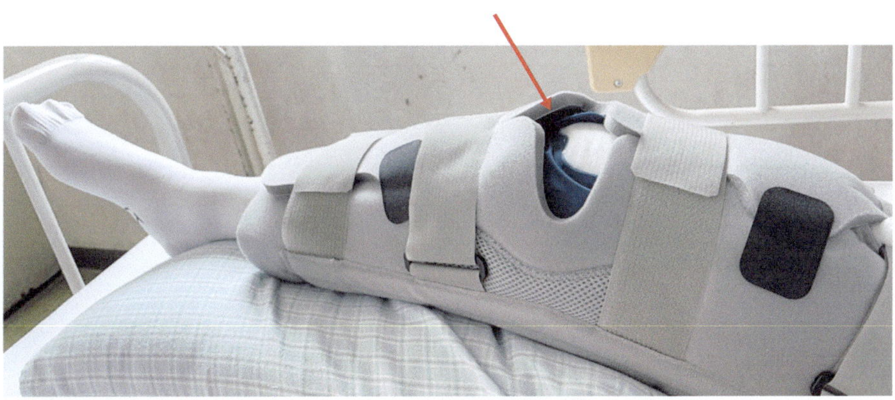

Fig. 9.3 Cylindrical soft brace for 1–2 week immobilization after surgery
Use in mild flexion (10–20°). For the first 3 days, cooling and compression are applied using a Cryo-cuff® (arrow)

tendon-bone graft fixed with an interference screw to the femoral tunnel, 1-week immobilization is recommended. For a case using a soft tissue graft such as a hamstring tendon graft fixed with pullout sutures on the femoral cortex, 2-week immobilization is recommended as the graft moves back and forth inside the femoral tunnel with flexion-extension movement. As to weight bearing, From the viewpoint of protection of the graft, partial weight bearing is permitted at 2–3 weeks, followed by full weight bearing at 3–5 weeks postoperatively.

POINT
During this period, maximum emphasis is on improving the range of motion, while minimizing stress on the tunnel wall and the graft so as not to interfere with the graft-tunnel integration. Muscle strengthening should be performed sparingly.

1. Management of postoperative inflammation

Cooling/compression devices including Cryo-cuff® are used to control joint edema and pain for 3 days after surgery during the period of immobilization.

Afterwards, anti-inflammatory physical therapies such as RICE (Rest, Icing, Compression, Elevation) treatment are performed. Electrical stimulation therapy is also effective. When patients begin full weight-bearing or walking, the inflammatory signs should be checked. They are instructed to control the amount of walking and to cool down with icing after exercise (for about 20 min).

2. Restoration of range of motion and flexibility

(1) Prevention of adhesions around the patella (Fig. 9.4)

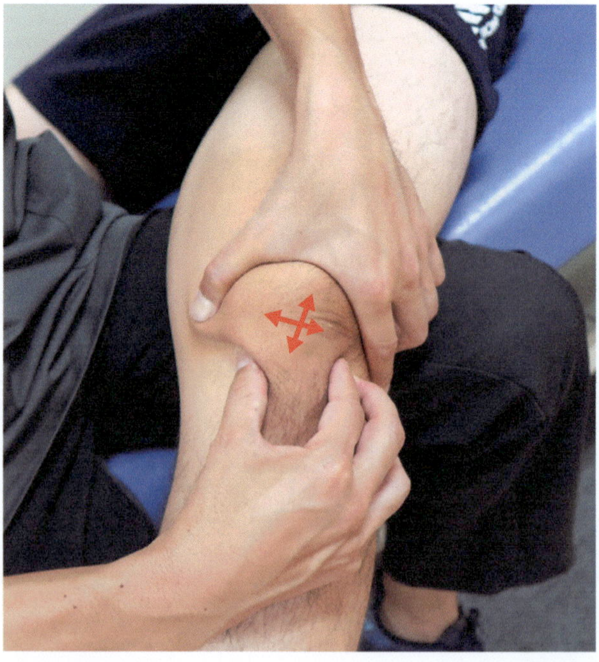

Fig. 9.4 Maneuver to prevent adhesion around the patella and the supra-patellar pouch Squeeze the patella and its surrounding soft tissue in proximal-distal, medially-laterally

Manual mobilization/stretching of the patella and surrounding tissues is performed to prevent adhesions and maintain flexibility as much as possible during the 1–2 week postoperative immobilization period.

(2) Improvement of soft tissue gliding in the distal thigh (Fig. 9.4)

Squeeze the soft tissues of the distal thigh/patella to improve sliding between the suprapatellar pouch, vastus medialis, and patellar retinaculum.

(3) Improvement of gliding of the infra-patellar fat pad (Fig. 9.4)

The infra-patellar fat pad is prone to scarring due to surgical manipulation through the anterior arthroscopic portals [4]. As scarring leads to adhesions of the surrounding tissues, manipulation from lateral to medial side is performed to improve its gliding.

(4) Quadriceps stretching

After starting range-of-motion exercise, the knee joint is flexed in hip extension (Fig. 9.5a). This method is useful to stretch the proximal portion of the rectus femoris muscle at a relatively small knee flexion angle [5] in the early postoperative period.

If the knee flexion angle exceeds 120°, the knee joint is flexed to the predetermined angle while the pelvis is held in a posterior tilt position, and then the hip is extended (Fig. 9.5b). This makes it possible to efficiently stretch the distal part of

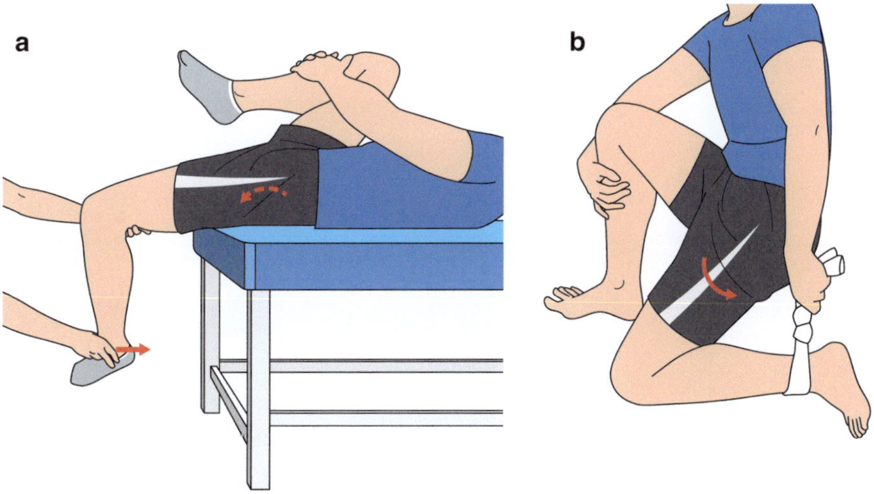

Fig. 9.5 Stretching of the quadriceps femoris muscle
(**a**) Stretching of the proximal quadriceps muscle in the modified Thomas test position: flex the knee (arrow) with the hip extended (dashed arrow)
(**b**) Stretching of the distal/lateral quadriceps: extend the hip after knee flexion in posterior pelvic tilt position (arrow)

the rectus femoris and vastus lateralis muscles without hyperflexion of the knee joint [6].

3. Improvement of muscle atrophy
(1) Quadriceps strengthening

Immediately postoperatively, the patient is instructed to perform the quadriceps setting and the straight leg raising (SLR) exercises with the brace [7]. Avoid vigorous contraction of the quadriceps in extension in the open kinetic chain (OKC) or posterior tibial tilting in the closed kinetic chain (CKC), as they produce posterior displacement of the femoral condyles [8, 9].

OKC training after the immobilization period is limited only in flexion over the quadriceps neutral angle (70°) with the resistance placed on the proximal tibia. (Fig. 9.6a).

(2) Knee flexor/hamstring muscle strengthening

Knee flexor/hamstring muscle training could be started immediately after ACL reconstruction using BTB graft [10]. On the other hand, after the reconstruction using hamstring tendon graft [11] concentric contraction of the hamstring muscle with leg curl should be avoided, and a bridge movement in hip extension using gluteus maximus contraction is recommended. (Fig. 9.6b).

(3) Simultaneous strengthening of quadriceps/hamstring muscles

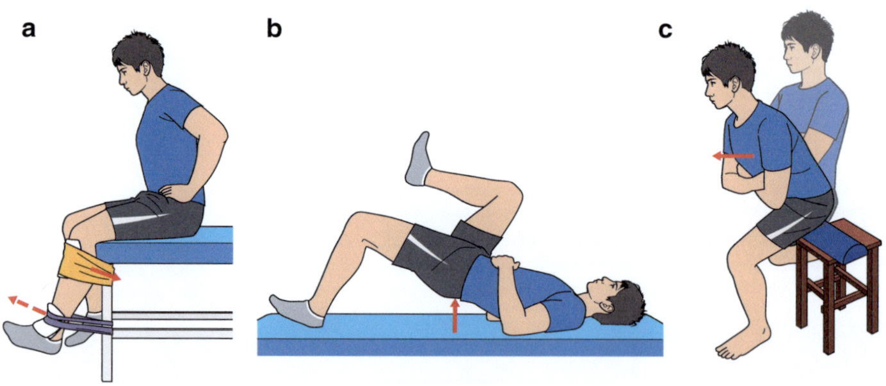

Fig. 9.6 Muscle strengthening training during the recovery period
(**a**) Knee extensors: Extend the knee (dashed arrow) while anterior tibial displacement is blocked (arrow).
(**b**) Hip extension exercise with co-contraction of the hamstring and the gluteus maximus muscles (bridge): Recommended for the hamstring muscle training after hamstring tendon graft harvesting
(**c**) Quadriceps muscle training by forward trunk movement (half sitting exercise): Perform in a half-sitting position while the ipsi-lateral buttock is placed on a table

In CKC exercises (e.g., squatting), simultaneous contraction of the quadriceps and hamstring muscles reduces posterior slipping the femoral condyles on the tibial plateau [12]. However, it must be performed with the lower leg tilted anteriorly to avoid posterior slippage of the femoral condyles on the tibial plateau [13] . In the half-sitting position, forward half sitting exercise (FHSE) is recommended (Fig. 9.6c), in which the affected lower limb is loaded anteriorly by the anterior leaning of the trunk. It can significantly increase quadriceps and hamstring activity while controlling abduction/adduction moments compared to squatting on both legs [14].

4. Weight bearing/walking

Weight bearing/walking should be started with 1/2 to 1/3 weight bearing at 2–3 weeks, and full weight bearing at 3–5 weeks. To protect the reconstructed graft, anterior tilt of the lower leg should be maintained, and toe-touch gait should be ensured [13]. A soft brace as shown in Fig. 9.7 should be worn during weight-bearing/walking to guide the patient to toe-touch gait (up to 3 months postoperatively).

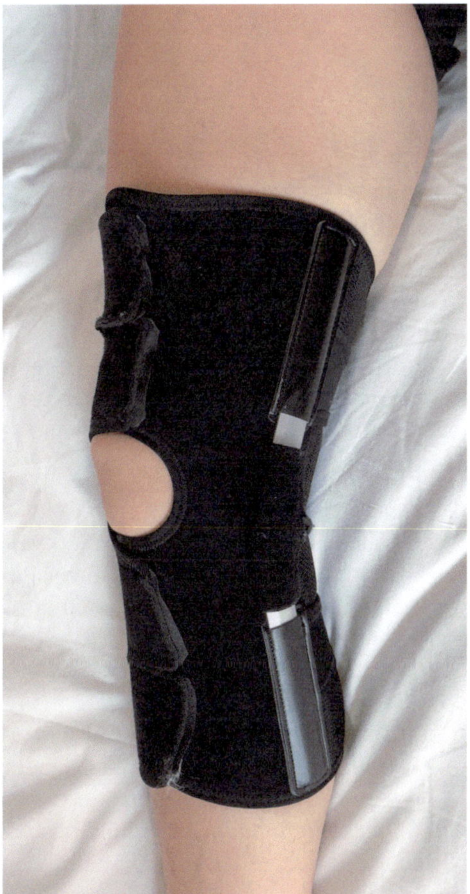

Fig. 9.7 A soft knee brace to protect the ACL graft at the time of walking. Full extension at the time of heel contact during walking is avoided and the patient is guided to toe contact instead. This prevents posterior slip of the femoral condyles due to the posterior tilt of the tibial plateau. This functional brace is recommended to wear for 3 months after surgery

POINTS

1. In the recovery period, the remodeling process of the reconstructed graft should be kept in mind. Special care should be taken not to overload the graft during the necrotic phase (approximately 2 months after surgery) when the graft is assumed weakest.
2. Joint contractures due to arthrofibrosis should be avoided at all costs. Stiff knees are worse than unstable knees. Risk factors for arthrofibrosis include acute surgery, women, concomitant meniscus suture, and medial collateral ligament injury/reconstruction. Special attention should be paid to treat patients with these conditions.
3. Postural control muscles are important to reduce loading on the reconstructed grafts and to prevent re-injury. The muscles of the trunk and those around the hip joint should always be trained.
4. At the time of weight bearing, it is important to distinguish between heel-contact gait in knee extension and toe-touch gait in slight flexion. The former increases the posterior tilt of the tibial plateau leading to posterior slippage of the femoral condyles and stress to the reconstructed grafts. The latter is safer for the grafts because the condyles slip anteriorly.

9.3 Early Training Period [6 Weeks to 3 Months Postoperatively (Fig. 9.8)]

By this period, the graft-tunnel wall integration is almost complete, and the graft itself is considered to be in the early stage of remodeling [1, 2]. The load to the reconstructed graft is progressively increased. Muscles are trained up to about 50% of maximal muscle strength for recovery to a running level. Bicycle riding to facilitate anterior tilt of the lower leg [14] and swimming are encouraged.

1. Muscle strengthening

(1) Quadriceps muscle

- The leaf spring exercise (Fig. 9.9a) is a kind of OKC exercise to strengthen the quadriceps muscles in extension [15].
- Squatting: Once squatting on both legs is achieved, the load on the affected leg is gradually increased to a one-leg squatting (Fig. 9.9b, c). Bulgarian squatting or forward lunge, in which the amount of load on the affected leg can be adjusted, may be used (Fig. 9.9b, c). When forward lunge is performed, heel-ground contact with a wide stride may cause a posterior inclination of the tibial plateau, leading to posterior slippage of the femoral condyles /anterior subluxation of the tibia [16]. For this reason, care should be taken to maintain ground contact from the forefoot with a narrow stride.

medical / athletic		medical rehabilitation			
elapsed postoperative period		6 weeks	8 weeks	10 weeks	3 months
rehabilitation category			first semester of training		
ROM (extension/flexion) 0/135°		Passive patellar manipulation → (0/135°) →			
Weight bearing/WB					
Muscle strength (OKC)	Quad.	leg extension (90° to 60° isotonicity) →	leaf spring exercise →		
	Ham.	leg curl (isotonicity) for HT →			
Muscle strength (CKC)		Bulgarian squat Forward lunge Deadlift	Squat (one leg)		Split squat Romanian deadlift Kettlebell swing
Endurance power		Stationary bicycle (no load) Swimming (other than breaststroke)	(Progressive loading) →		
Posture control			Disturbance on bipedal support →		Disturbance one-legged support Core muscle training (CKC)
Shock absorption			Modified drop squat (both legs)	Modified drop squat (one leg)	Mini-jump on both legs
Instantaneous power					
Sprint					Running (jogging)
Agility					
Competition					

HT: Reconstruction with hamstring tendon graft ; OKC: open kinetic chain; CKC: closed kinetic chain; Quad.: quadricep femoris muscle; Ham.: hamstring muscle

Fig. 9.8 Phase II: Rehabilitation program in the first semester of training (6 weeks to 3 months postoperatively)

Resistive lateral leg reach (Fig. 9.9d), in which the contralateral leg reaches laterally against resistance during a one-legged squatting, increases hip abduction moment and suppresses knee abduction moment [17].

(2) Hamstring muscles

Hamstring muscle strengthening after the reconstruction using the hamstring tendon is performed with CKC training, while restricting hip flexion as in the dead lift. From 8 weeks postoperatively, leg curl should be started with a low load, and then with the load gradually increased.

2. Postural control

Postures at the time of ACL injury include shallow flexion angle of the knee joint [18], excessive abduction and rotation of the knee joint, increased internal rotation of the hip joint, and backward tilt of the trunk [19–22]. It should be aimed to improve postural control functions to avoid these risky postures.

- Compared to both-leg support, one-leg squatting increases the lower limb muscle torque required for weight bearing and decreases the ground contact area of the foot, which may lead to poor posture. In the first half of the training period, the subject is instructed to keep squatting posture with the knee joints in mid-

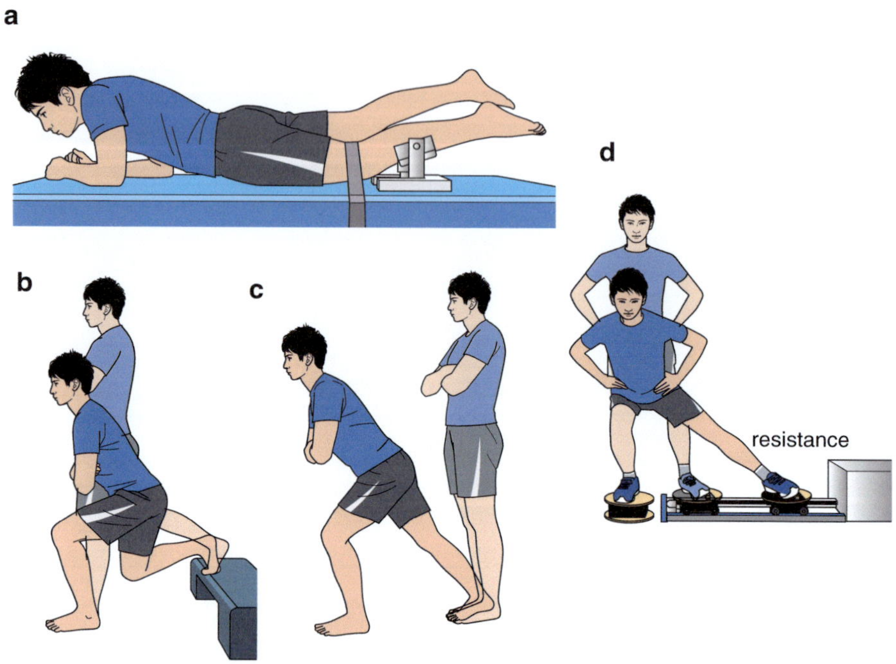

Fig. 9.9 Strength training in the early training period
(**a**) Leaf spring exercise for quadriceps strengthening in extension: The anterior aspect of the proximal calf is used as the fulcrum of support, and resistance is applied to the posterior aspect of the distal thigh
(**b**) Bulgarian squat: load is adjusted by placing the healthy lower limb on a posterior platform
(**c**) Forward lunge: hold the lower legs and trunk in a forward leaning position
(**d**) Resistive lateral leg reach: reach the contralateral lower leg against resistance during one-legged squat

flexion on both legs, and then transfer to one-leg support on the affected leg by releasing the healthy leg as gently as possible without moving the trunk and pelvis. Postural dyscontrol can be easily detected by the compensatory movements that occur during weight transfer. Patients are readily aware of the compensatory movements, and can be trained to acquire good posture control [23] (Fig. 9.10).

- Many patients with knee-abduction (knee-in) have excessive hip adduction and internal rotation. In these cases, Romanian dead lift by one leg is performed to promote hip abduction and external rotation, and to control rotation of the pelvis (Fig. 9.11a). In cases with foot pronation, the patient's foot is stabilized by loading the great toe ball, little toe ball and heel, which are the base points of the plantar arch (Fig. 9.11b).
- The lower limbs are held stationary in a posture with the legs opened back and forth. The postural control functions are improved by applying disturbance through trunk and upper limbs. (Fig. 9.11c).

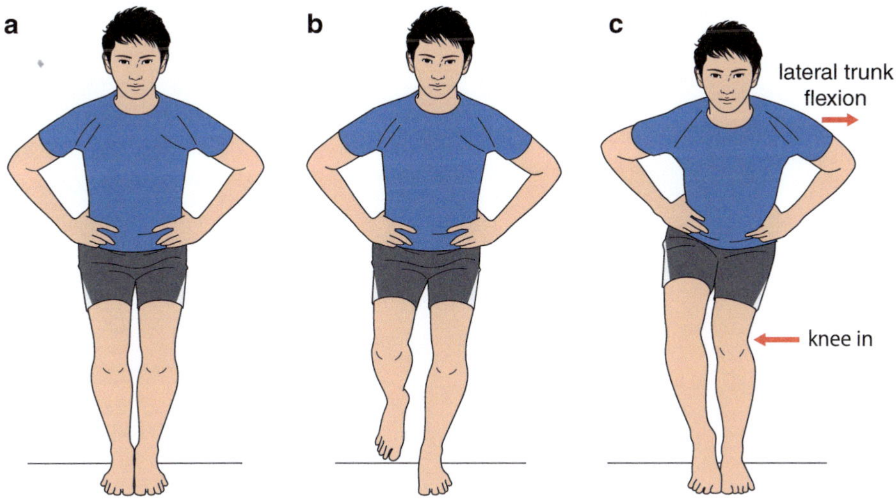

Fig. 9.10 Postural control evaluation during one-leg squatting
(**a**) Starting limb position: hold the knee joint in mid-flexion
(**b**) One-leg supported posture: A good example of one-leg supported posture without compensatory movement of the trunk and pelvic girdle
(**c**) Compensatory posture: A defective case in which a compensatory movement occurs

3. Improvement of shock-absorbing function

- The modified drop squat (MDS) [24, 25] (Fig. 9.12), in which the subject descends from a toe-up position and stops by flexing the knee joint at the moment of heel contact, is performed from both legs. The MDS with one leg, which is performed in a stepwise fashion, places less load on the knee joint than during running or ground contact at jumping, and can improve the shock-absorbing function of the lower limb including the knee joint, while avoiding shock load. This restores the function of the quadriceps muscle and the patellar tendon as a spring to release centrifugal energy.
- The speed of one-leg calf raise is gradually increased to perform reactive ankle joint exercises to restore the shock-absorbing function of muscle-tendon complex around the ankle joint.

POINT

1. Muscle strengthening and postural control gradually shift from bilateral to single leg support.
2. Thoroughly improve the postural control function by one-leg squatting.
3. The shock-absorbing function of the knee joint is performed with a modified drop squatting.

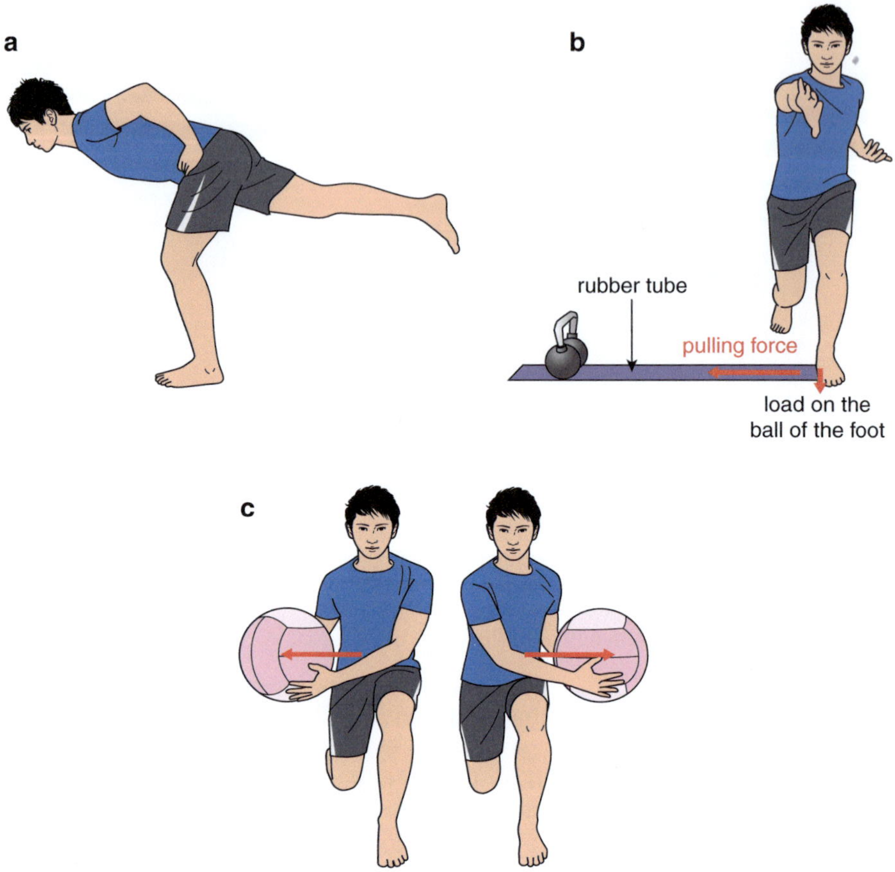

Fig. 9.11 Posture control training. (**a**) Romanian dead lift: Promotes hip abductor and external rotator activity while controlling pelvic rotation. (**b**) Load shift to the great toe ball of the foot: While loading on the great toe ball, hold the rubber tube so that it does not come loose, and rotate the upper body. (**c**) Disturbance application: Disturbance caused by movement of the trunk or upper limbs while the lower limbs are stabilized

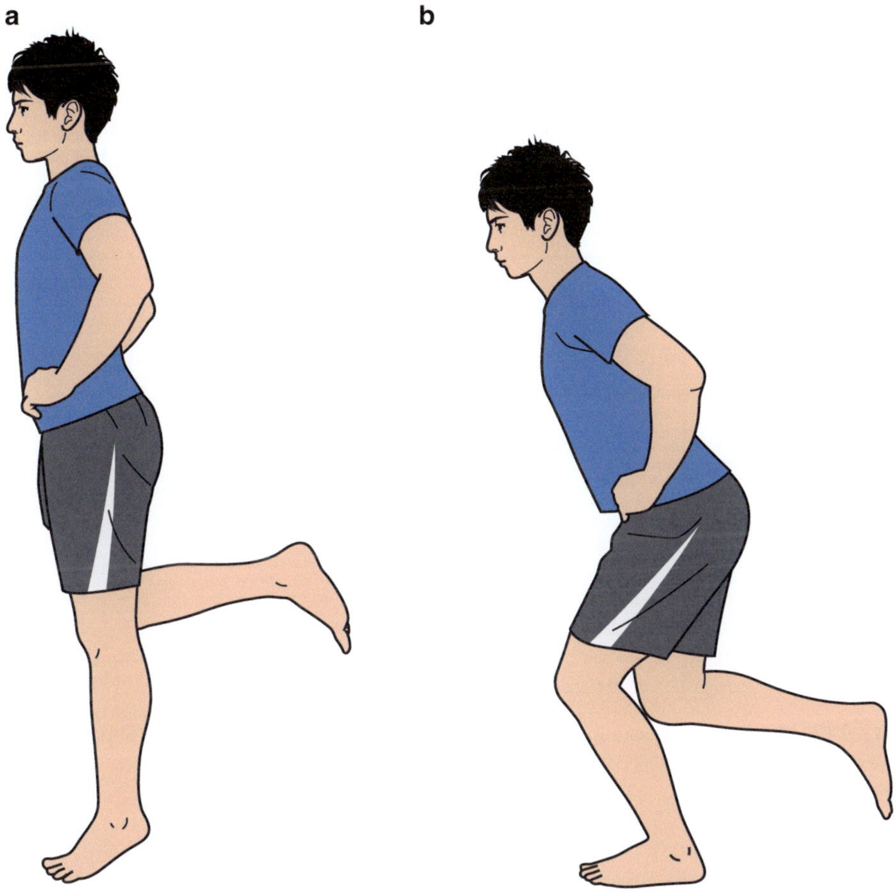

Fig. 9.12 Modified drop squatting (MDS) (**a**) Starting position: Standing on tiptoes with lower limbs and trunk held straight. (**b**) End position: Descending with weakness, shock absorption is achieved by heel contact during knee flexion

9.4 Late Stage of Training [3–6 Months Postoperatively (Fig. 9.13)]

This is the mid-stage during remodeling of the graft [26, 27]. It is time to further increase the load on the graft [26] (up to about 90% of maximal exercise intensity) and to bring the muscle recovery to the level of return to sports. Specifically, the patients will reacquire high-intensity exercises such as sprinting, jumping, and change of direction, and begin practicing basic event-specific skills.

1. Muscle strengthening: Aiming to reduce the injured-normal difference.

Medical/Athletic		Athletic rehabilitation				
elapsed postoperative period		4 months	5 months	6 months	7 months	9 months
Rehabilitation Category		Late stage of training			return-to-sports period	
ROM (extension/flexion)		0/135° (0/135°) →	0/140° (0°) →	kneeling with the knees fully flexed, and with dorsum of the feet kept on the floor		
Weight bearing/WB						
Muscle strength (OKC)	Quad.	leg extension(isokinetic) (high speed - recoil) →		- - - - →		
		leaf spring exercise →		- - - - →		
	Ham.	leg curl(isokinetic) →		- - - - →		
Closing power (CKC)		Weight-loaded →				
Endurance		Fixed bicycle (load increment) Endurance running →				
Posture control		On unstable surfaces (Whole body vibration equipment)	disturbance load: gradually increased →			
shock absorption		two-legged jump One leg mini-jump	One leg jumping/hopping One leg jump - landing	→		
instantaneous force		power clean /box jump (both legs) reactionary rise rope-jumping Bicycle power training	box jump (one leg) depth jump continuous jump →			
sprint		running push lunge	sprint →			
agility		lateral squat	Agility (ladder) cutting Stop, turn	→		
competition			Event-specific skills	athletic action (with vs. without opponents)	practice match	return-to-sports / competition

Fig. 9.13 Phase III-IV: Rehabilitation program from the late stage of training (3–6 months after surgery) to the return-to-sports period (6 months after surgery or later)

In OKC, the leg extension load is progressively increased. In CKC, a split squat is useful in which the lower leg of the healthy limb, the trunk, and the thigh of the affected leg are kept vertical (Fig. 9.14a). This exercise is considered to be safe for the graft, as the femoral condyle slides forward while quadriceps activity is increased [28].

Contraction of the semitendinosus muscle compresses the medial compartment of the knee joint and reduces dynamic knee abduction. Especially Romanian dead lift (Fig. 9.14b) and kettlebell swing (Fig. 9.14c) increase the semitendinosus muscle activity rather than the biceps femoris [29].

The squatting on both legs should be performed with a weight below body weight, paying attention not to lead to compensatory movements.

2. Shock absorption and postural control
(1) Neuromuscular training

Neuromuscular training that provides disturbances that disrupt postural stability, improves motion patterns to reduce the risk of ACL injury [30].

First, one-leg squats (static and dynamic) are performed with active trunk and pelvis rotation or loading of the healthy lower limb, both upper limbs, and the ankle

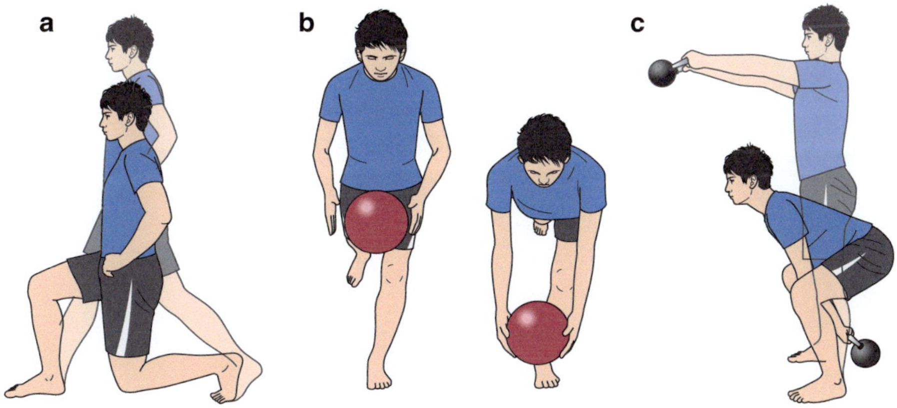

Fig. 9.14 Strength training in the late phase of training
(**a**) Split squat: During descending movement, keep the lower leg of the healthy limb, the trunk, and the thigh of the affected leg vertical
(**b**) Romanian dead lift: Catch the medicine ball quickly after once releasing it
(**c**) Kettlebell swing: Swing the kettlebell by flexion and extension movements of the hip joint

joint in the affected limb to improve postural control (Fig. 9.15a, b). Next, devices to sway center-of-gravity, such as unstable surfaces, water bags or whole body vibration device (after 4 months postoperatively), are used (Fig. 9.15c–e).

(2) Shock absorption training

The joints are subjected to strong impact loading during sprinting, jumping, and change of direction, which are the potential triggers of ACL injury. Thus improvement of the shock-absorbing function is important for preventing re-injury. It was reported that the knee flexion angle during sprinting in this period decreased compared to the control group [31], suggesting insufficient shock absorption due to decreased knee flexion angle during the stance phase. The difference in knee flexion angle was correlated with the difference in knee flexion angular velocity in the one-leg MDS [32]. In other words, the acquisition of smooth knee flexion and extension during one-leg MDS is important for the acquisition of symmetrical sprinting.

The landing impact of a jump is smaller when the time between ground contact and the peak of floor reaction force is longer, and center-of-gravity sway can be suppressed by mitigating the impact [33]. Mini-jumps, in which the athlete stabilizes the forefoot by simultaneously touching the great toe ball and the small toe ball at the moment of contact without touching the heel, are useful for strengthening the shock-absorbing function (Fig. 9.16a). In addition, moderate anterior tilt of the trunk and pelvis places the center of foot pressure forward and increases plantar flexion and hip extension moments, leading to reduction of impact to the knee joint [34, 35]. To stimulate trunk muscle contraction and strengthen forefoot support in preparation for landing impact, one-leg jumps from a position in which the trunk is

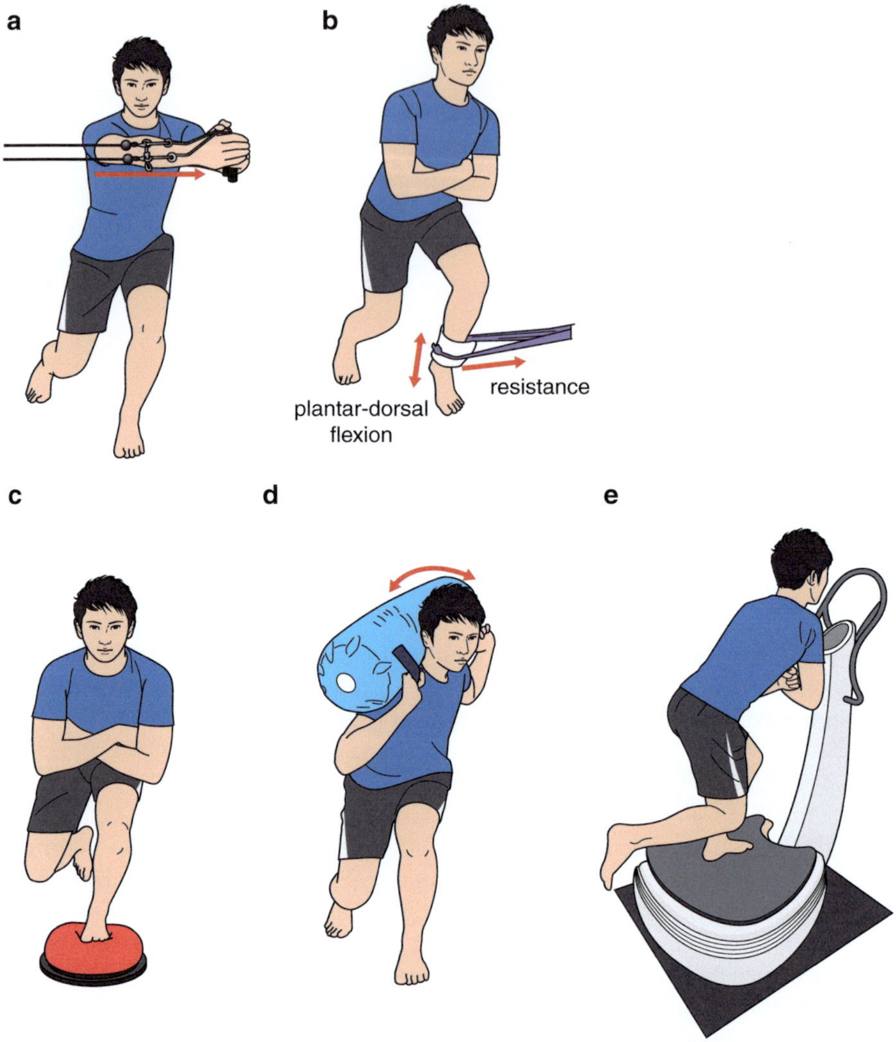

Fig. 9.15 Neuromuscular training
(**a**) Upper limb loading: Horizontal resistance is applied to the upper limb to stabilize the lower body (contralateral lower limb is lightly loaded)
(**b**) Ankle joint loading: Stabilize the foot by applying medial and lateral loading to the ankle joint in the affected side
(**c**) Unstable Surfaces: Using devices to induce multi-directional instability
(**d**) Water bag loading: Disturbance load is applied by movement of water in the bag (contralateral lower extremity is lightly loaded)
(**e**) Whole body vibration device: Disturbance loading with a device that generates three-dimensional vibration (Power Plate®; Performance Health System, Inc.)

Fig. 9.16 Landing training
(**a**) Forefoot: The big toe and small toe balls are grounded simultaneously to stabilize the forefoot transverse arch.
(**b**) Two hands and one leg support: Spine and contralateral lower limb are held horizontally, and the supporting leg lands and rises with its forefoot

held horizontally and supported by both hands and one foot, are recommended (Fig. 9.16b). Once the one-leg drop-jump landing and single leg hop are stabilized, high power plyometric trainings with strong impacts such as tuck jumps, depth jumps in each direction, continuous jumps using mini hurdles, and continuous hopping, are performed to improve the spring function, postural and impact control of one leg, leading to a competitive level.

3. Recovery of acceleration/deceleration and agility

(1) Leap

After 4 months postoperatively, power clean (Fig. 9.17a) and box jump (jump on a platform), which emphasize loading of the affected lower limb, should be performed as lower limb joint strength training. For hip strength training, increase the load and speed of the afore-mentioned Romanian dead lit and kettlebell swing. For knee joint strength training, leg extension at high speed in OKC and a reactionary standing up exercise in CKC are performed (Fig. 9.17b). For ankle joint strength training [36–38], mini-jumps and rope-jumping are performed with the knee mildly flexed.

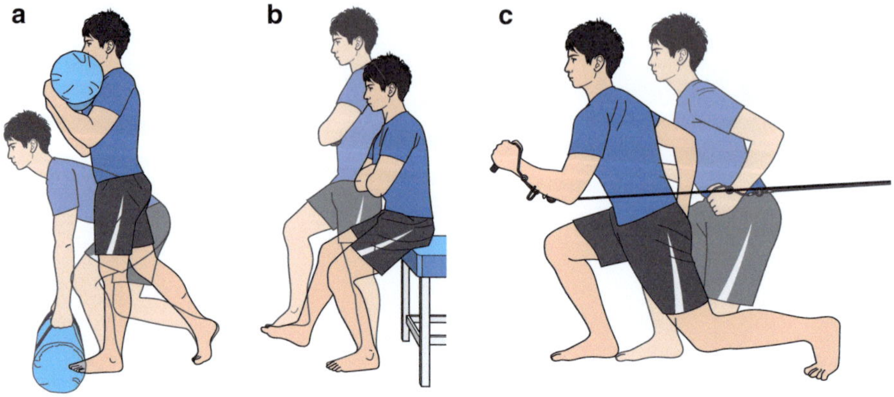

Fig. 9.17 Joint strength training
(**a**) power clean: From a posture in which the load on the affected lower limb is emphasized, the weight is raised by rapid extension of the index lower limb.
(**b**) Rise from sitting: The knee joint is flexed and extended with rapid relaxing
(**c**) Push lunge (anterior movement by hip extension of the posterior leg): Cultivate hip extension power while maintaining knee flexion

(2) Running

The speed of running is gradually increased to 50% of full speed by 5 months postoperatively. In order to improve the deceleration as well as the acceleration function, the patients practice decelerating and stopping in a straight line controlling stride length. After 5 months postoperatively, the patients begin sprinting in a straight line, and gradually increase speed to full. At the same time, the number of steps taken during deceleration is gradually decreased to restore the ability of sudden stop in a straight line, leading to turn and cutting from the sudden stop. In sprinting, the hip joint is extended with the knee extension controlled in the latter half of the kicking motion, effectively converting the hip joint extension velocity into a backward swing velocity of the entire leg [39]. In the afore-mentioned split squat posture, the push lunge exercise is performed in which the hip joint of the affected side is quickly extended from the flexed position to move the trunk forward while leaving the knee joint behind and inhibiting the upward shift of the center of gravity (Fig. 9.17c).

(3) Cutting

Sidestepping is essential for athletic performance, but it may cause ACL injury. To speed up the movement, it is necessary to shorten the plantar contact time, to strengthen hip extension power for deceleration, and to power up plantar flexion for acceleration [40, 41]. Begin with a safe stride length and speed, as large stride length and/or increased speed could be risky. The trunk should be held in the mid-position, grounding from the heel should be avoided, while grounding from the forefoot is recommended [42, 43].

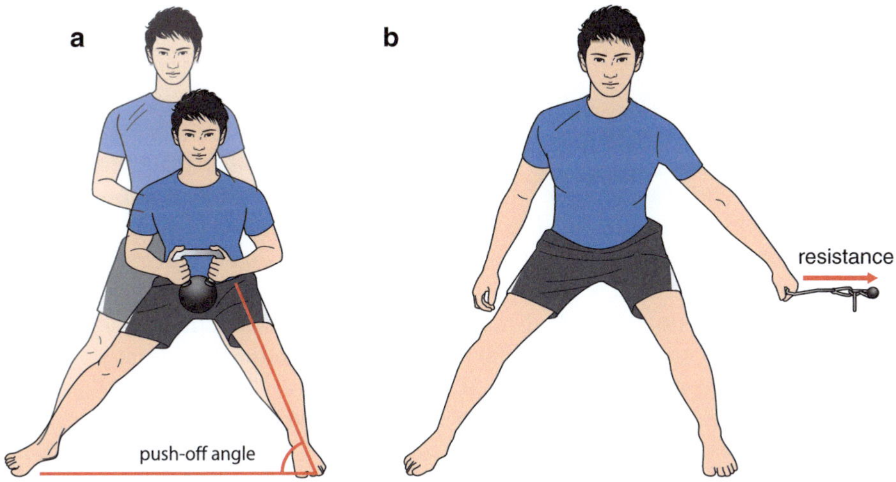

Fig. 9.18 Lateral squatting
(**a**) The lower limb joints are flexed and extended while maintaining the push-off angle. This shifts the body's center of gravity laterally
(**b**) Resistive load is applied from the side to increase acceleration, leading to a cutting motion

The lateral squat is performed as one of the cutting exercise trainings [44] (Fig. 9.18a). The hip, knee, and ankle joints are maintained in the intermediate positions in rotation with the trunk in the middle position, and the lower limbs are flexed and extended while maintaining the push-off angle. This causes a lateral shift of the center of gravity. In this exercise, it is important to avoid rotational loading of the knee joint by aligning the centers of all of the joints in the lower limb. Once the lateral squat is achieved, the subject can shift to a reactionary bouncing motion of repeatedly pushing off and landing on the forefoot in the same posture. Then add resistance from the lateral side to improve the movement speed (Fig. 9.18b).

POINT

1. Dynamically reduce the risk to avoid knee joint abduction by strengthening the medial hamstring muscles.
2. Neuromuscular training is performed to obtain good postural control.
3. After acquiring the shock-absorbing function, the instantaneous force/acceleration is gradually increased.
4. Improve acceleration and deceleration functions to achieve both agility and safety in each movement.

9.5 Return-to-Sports Period [6 Months After Surgery ~ (Fig. 9.13)]

The maturation of the graft might take several years [1], and the maximum failure load of the graft during this period could be assumed smaller than that of the normal ACL [45, 46]. Therefore, the prevention of re-injury should be kept in mind, especially during the early period of return-to-sports. Always try to further improve postural control and shock absorption. Improve each movement and event-specific skill to a competitive level.

Since there are no established criteria for return to competition, patients are allowed to return to competition, taking into consideration the joint condition including stability, muscle strength, and athletic function. The followings are suggested guidelines for return to competitive practice: (1) no limitation of joint range of motion; (2) no detectable joint instability; (3) 90% or greater in injured/non-injured ratio of thigh muscle strength ; (4) ability to perform acceleration/deceleration, jump landing, and change of direction without escape posture; (5) ability to perform competition-specific skills other than interpersonal play. Once these are achieved, the patient will begin competition-practice avoiding interpersonal play, with the aim of a full return by 9 months postoperatively, while being advised to keep training based on physical findings and performance status.

References

1. Kinugasa K, Shino K, et al. Cross-sectional area of hamstring tendon autograft after anatomic triple-bundle ACL reconstruction. Knee Surg Sports Traumatol Arthrosc. 2017;25:1219–26.
2. Kinugasa K, Shino K, et al. Chronological changes in the cross-sectional area of the bone-patellar tendon-bone autograft after anatomic rectangular tunnel ACL reconstruction. Knee Surg Sports Traumatol Arthrosc. 2021;29:3782–92.
3. Shelbourne KD, et al. Loss of normal knee motion after anterior cruciate ligament reconstruction is associated with radiographic arthritic changes. Am J Sports Med. 2012;40:108–13.
4. Tang G, et al. Fibrous sear in the infrapatellar fat pad after arthroscopy: MR imaging. Radiat Med. 2000;18:1–5.
5. Kimura Y, et al. Relationship between the elongation limb position and tissue elasticity of the quadriceps muscle – comparison between the Modified Thomas test limb position and the Ely test limb position. J Jpn Soc Orthopedic Ultrasonics. 2018;29:38–44. (in Japanese)
6. Kimura Y, et al. Relationship between quadriceps muscle elongation method and tissue elasticity – influence by pelvic limb position. J Jpn Soc Orthopedic Ultrasonics. 2017;28:28–33. (in Japanese)
7. Mae T, Shino K, et al. Graft tension during active knee extension exercise in anatomic doble-bundle anterior cruciate ligament reconstruction. Arthroscopy. 2010;26:214–22.
8. Boden BP, et al. Tibiofemoral alignment: contributing factors to noncontact anterior cruciate ligament injury. J Bone Joint Surg Am. 2009;91:2381–9.
9. Daniel DM, et al. Use of the quadriceps active test to diagnose posterior cruciate-ligament disruption and measure posterior laxity of the knee. J Bone Joint Surg Am. 1988;70:386–91.
10. Shino K, et al. Rectangular tunnel double-bundle anterior cruciate ligament reconstruction with bone-patellar tendon-bone graft to mimic natural fiber arrangement. Arthroscopy. 2008;24:1178–83.

11. Shino K, et al. Anatomic anterior cruciate ligament reconstruction using two double-looped hamstring tendon grafts via twin femoral and triple tibial tunnels. Oper. Tech Orthop. 2005;15:130–4.
12. Palmitier RA, et al. Kinetic chain exercise in knee rehabilitation. Sports Med. 1991;11:402–13.
13. Koyanagi M, Shino K, et al. Effets of changes in skiing posture on the kinetics of the knee joint. Knee Surg Sports Traumatol Arthrosc. 2006;14:88–93.
14. Tada S, et al. Kinematic and electromyographic validation of lower limb muscle training by trunk forward leaning in Half sitting. Phys Ther. 2019;46:233–41. (in Japanese)
15. Nakae N, et al. Effect of quadriceps training with a lower leg support bridge in the abdominal position on anterior tibial translation in anterior cruciate ligament insufficiency knees. Clin Biomech. 2009;30:425–30. (in Japanese)
16. Kimura Y, et al. Analysis of foreleg motion in forward lunge. Jpn J Clin Biomech. 2004;25:425–9. (in Japanese)
17. Kimura Y, et al. Kinematic analysis of the supporting leg in lateral resisted leg reaching motion. Clin Biomech. 2010;31:445–52. (in Japanese)
18. Olsen OE, et al. Injury mechanisms for anterior cruciate ligament injuries in team handball: a systematic video analysis. Am J Sports Med. 2004;32:1002–12.
19. Koga H, et al. Mechanisms for noncontact anterior cruciate ligament injuries: knee joint kinematics in 10 injury situations from female team handball. Am J Sports Med. 2010;38:2218–25.
20. Krosshaug T, et al. Mechanisms of anterior cruciate ligament injury in basketball: video analysis of 39 cases. Am J Sports Med. 2007;35:359–67.
21. Boden BP, et al. Video analysis of anterior cruciate ligament injury: abnormalities in hip and ankle kinematics. Am J Sports Med. 2009;37:252–9.
22. Hewett TE, et al. Video analysis of trunk and knee motion during noncontact anterior cruciate ligament injury in female athletes: lateral trunk and knee abduction motion are combined components of the injury mechanism. Br J Sports Med. 2009;43:417–22.
23. Kimura Y, et al. Rehabilitation after meniscus suture. Athletic rehabilitation guide. In: Fukubayashi T, et al., editors. Practical approach for return to competition and prevention of recurrence. 2nd ed. Tokyo: Bunkodo; 2018. p. 210–7. (in Japanese).
24. Kondo S, et al. Kinematic analysis of the modified drop squat as an assessment of shock absorption function. Clin Biomech. 2016;37:327–34. (in Japanese)
25. Kimura Y, et al. Kinematic analysis of modified drop squat – dynamic evaluation of shock absorption function. Sports Injury. 2016;21:24–6. (in Japanese)
26. Goradia VK, et al. Tendon-to-bone healing of a semitendinosus tendon autograft used for ACL reconstruction in a sheep model. Am J Knee Surg. 2000;13:143–51.
27. Ménétrey J, et al. "Biological failure" of the anterior cruciate ligament graft. Knee Surg Sports Traumatol Arthrosc. 2008;16:224–31.
28. Kimura Y, et al. Kinematic analysis of the split squat. Clin Biomech. 2011;32:441–8. (in Japanese)
29. Buchanan TS, et al. Selective muscle activation following rapid varus/valgus perturbations at the knee. Med Sci Sports Exerc. 1996;28:870–6.
30. Hewett TE, et al. Anterior cruciate ligament injuries in female athletes: part 2, a meta-analysis of neuromuscular interventions aimed at injury prevention. Am J Sports Med. 2006;34:490–8.
31. Lewek M, et al. The effect of insufficient quadriceps strength on gait after anterior cruciate ligament reconstruction. Clin Biomech. 2002;17:56–63.
32. Kimura Y, et al. Association between modified drop squat and kinematic index in jogging after knee anterior cruciate ligament reconstruction surgery. Jpn J Clin Sports Med. 2018;26:S253. (in Japanese)
33. Kimura Y, et al. Dynamic balance analysis of drop-jump landings: relationship between shock absorption after landing and center of gravity sway. Jpn J Clin Sports Med. 2013;23:45–6. (in Japanese)

34. Shimokochi Y, et al. The relationships among sagittal-plane lower extremity moments: implications for landing strategy in anterior cruciate ligament injury prevention. J Arhl Train. 2009;44:33–8.
35. Shimokochi Y, et al. Changing sagittal plane body position during single-leg landings influences the risk of non-contact anterior cruciate ligament injury. Knee Surg Sports Traumatol Arthrosc. 2013;21:888–97.
36. Marshall BM, et al. Biomechanical factors associated with time to complete a change of direction cutting maneuver. J Strength Cond Res. 2014;28:2845–51.
37. Debaere S et al. From block clearance to sprint running: characteristics underlying an effective transition. J Sports Sci. 2013;31:137–149.
38. Vanezis A et al. A biomechanical analysis of good and poor performers of the vertical jump. Ergonomics. 2005;48:1594–603.
39. Ito A. Running motion and muscle activity of sprinters. In: Kaneko M, et al., editors. Biomechanics – scientific basis of physical exercise. Tokyo; 2004. p. 197–202. (in Japanese).
40. Dos' Santos T, et al. Mechanical determinants of faster change of directional speed performance in male athletes. J Strength Cond Res. 2017;31:696–705.
41. Havens KL, et al. Cutting mechanics: relation to performance and anterior cruciate ligament injury risk. Med Sci Sports Exerc. 2015;47:818–24.
42. Kristianslund E, et al. Sidestep cutting technique and knee abduction loading: implications for ACL prevention exercises. Br J Sports Med. 2014;48:779–83.
43. Dempsey AR, et al. Changing sidestep cutting technique reduces knee valgus loading. Am J Sports Med. 2009;37:2194–200.
44. Sato K, et al. The lateral squat. J Bodyw Mov Ther. 2013;17:560–2.
45. Pauzenberger L, et al. "Ligamentization" in hamstring tendon grafts after anterior cruciate ligament reconstruction: a systematic review of the literature and a glimpose into the future. Arthroscopy. 2013;29:1712–21.
46. Claes S, et al. The "ligamentization" process in anterior cruciate ligament reconstruction: what happens to the human graft? A systematic review of the literature. Am J Sports Med. 2011;39:2476–83.

Chapter 10
Outcome Evaluation of the ACL-Reconstructed Knee

The clinical outcome of the ACL-reconstructed knee is largely dependent on the condition of the reconstructed ligament as well as the articular cartilage, meniscus, and surrounding muscles, etc. Thus, the clinical outcome may be evaluated as good in the short term even if the reconstructed ligament is not functioning. Therefore, the evaluation of the ACL-reconstructed knee should be divided into two parts: 1. evaluation of the reconstructed ACL graft itself; 2. overall clinical evaluation of the ACL-reconstructed knee.

10.1 Evaluation of the Reconstructed ACL Itself

1. Physical findings

The goal of ACL reconstruction is, of course, to restore stability without loss of motion. Patients with insufficient restored stability or impaired range of motion should be categorized into "poor".

Currently, the following conditions are criteria on physical findings to determine the success of ligament reconstruction surgery.

1. Little or no limitation of joint range of motion (<5°)
2. Manual laxity test: Lachman test negative, pivot-shift test negative
3. Less than 2mm difference between the healthy and the affected side in measured anterior laxity using instruments including KT-1000, KT-2000, KSM--.

2. Imaging

Evaluation of the reconstructed ligament by physical examination is unreliable because it uses manual drawer/*in vitro* level force of 100–200 N. MRI of the reconstructed ligament in extension is recommended as a highly objective method. At the

K. Shino, *Essence of Anterior Cruciate Ligament*,
https://doi.org/10.1007/978-981-99-6536-6_10

same time, intra-articular tunnel apertures must be evaluated using 3-D CT, while a small amount of radiation exposure (about 0.08 mSv) is required. Non-anatomically located tunnel apertures should be considered "poor".

The normal ACL is observed as a linear tense shadow in extension (Fig. 5.11a).

A "good" reconstructed ligament is observed to be similar to the normal ligament in full extension. The following two points are necessary to determine "good" (Figs. 10.1 and 10.2) [1].

(1) Thickness, tension, and orientation of the reconstructive ligaments
    - On the oblique sagittal section, a tense linear ligamentous shadow with normal ligament-like thickness is observed along Blumensaat's line.
    - On the oblique coronal slice, a tense linear ligamentous shadow with normal ligament-like thickness is observed, which forms an angle of 70–75° to the tibial plateau.
(2) Femur-tibia positional relationship

The sagittal section shows no anterior tibial displacement and normal in the femur-tibia positional relationship. This is confirmed by the anterior margin of the reconstructed ligament/Blumensaat's line, observed in oblique sagittal slice as a tense linear ligament shadow, passing through Parsons' knob or the anterior border of the tibial tunnel correctly-created at the anatomical attachment area.

Fig. 10.1 An example of "good" ACL graft at 1 year after the anatomical rectangular tunnel reconstruction using the patellar tendon
(**a**) On the oblique sagittal MRI slice (T2*-weighted) in extension, a well-tensioned straight ligament shadow (arrow) with sufficient thickness is observed along Blumensaat's line (dashed line), and the anterior margin of the reconstructed ligament/Blumensaat's line is in line with the anterior border of the tibial tunnel.
(**b**) On the oblique coronal MRI slice (T2*-weighted) in extension, a tense linear ligamentous shadow (arrow) with sufficient thickness is also observed, which forms an angle of approximately 75° to the tibial plateau. A regenerated AML portion (dashed arrow) is observed. In the case with a patellar tendon with bone plugs, there is no anterior distal deviation of the graft within the femoral tunnel (*).
(**c**) Arthroscopic/intraoperative view of the reconstructed ligament with the patellar tendon graft. AML(Anteromedial-lateral) portion is absent (dashed arrow)
(**d**) Arthroscopic finding at 6 months postoperatively. Regenerated AML portion (dashed arrow) is observed
(**e**) 3-D CT image showing the tibial plateau at 3 weeks postoperatively. The rectangular tibial tunnel aperture (arrow) is desirably located within the anatomical attachment area.
(**f**) 3-D CT image showing the lateral wall of the intercondylar notch at 3 weeks postoperatively. The rectangular femoral tunnel aperture (arrow) is located within the anatomical attachment area

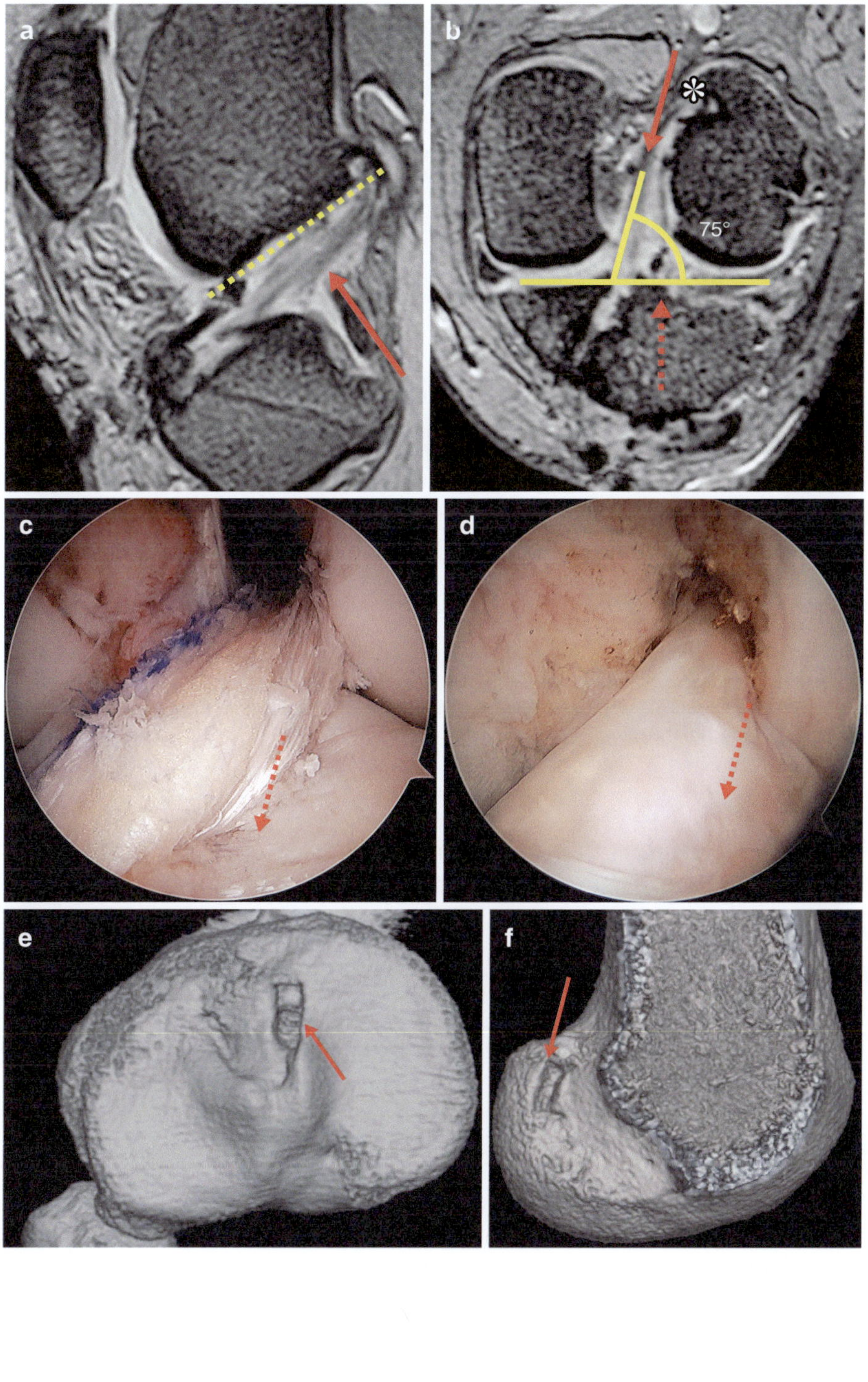
a
b
75°
c
d
e
f

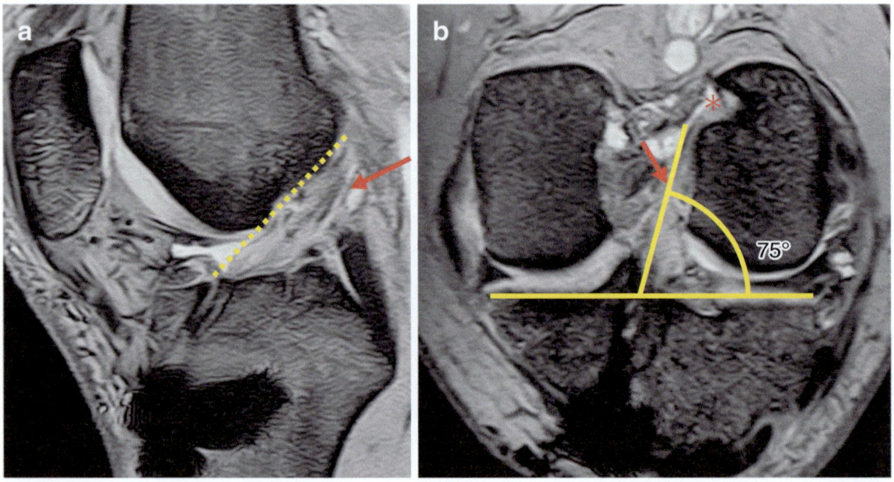

Fig. 10.2 An example of ACL graft at 1 year after the anatomical triple bundle reconstruction using the semitendinosus tendon
(**a**) On the oblique sagittal MRI slice (T2*-weighted) in extension, a well-tensioned straight shaded ligament of sufficient thickness (arrow) is observed along Blumensaat's line (dashed line). The anterior margin of the reconstructed ligament/Blumensaat's line coincides with the anterior border of the tibial tunnel aperture
(**b**) On the oblique coronal MRI slice (T2*-weighted) in extension, a well-tensioned, linear ligamentous shadow (arrow) is observed, which forms an angle of approximately 75° to the tibial plateau. In the case with soft-tissue grafts such as the semitendinosus tendon, the graft shifts anterodistally within the femoral tunnels (*)

10.2 Comprehensive Evaluation of the ACL-Reconstructed Knees

Only those with successful ACL grafts deserve this overall clinical evaluation while those with unsuccessful grafts do not deserve it (Fig. 10.3).

Various subjective/functional evaluation methods have been reported, but they tend to produce high scores and are not useful to reflect the comprehensive condition/function of the index knee [2]. The patient-oriented questionnaire [Japanese Anterior Cruciate Ligament Injury Questionnaire 25 (JACL-25) (Table 10.1)] is recommended as a subjective evaluation method that reflects function [3]. It is a 25-item questionnaire with a score ranging from 0/minimum to 100/maximum. Preoperative unstable knee cases score 70–80, and postoperative scores improve to about 20. A score of 20 or less indicates that the patient can engage in sports activities with little or no difficulty and can be categorized into "good results".

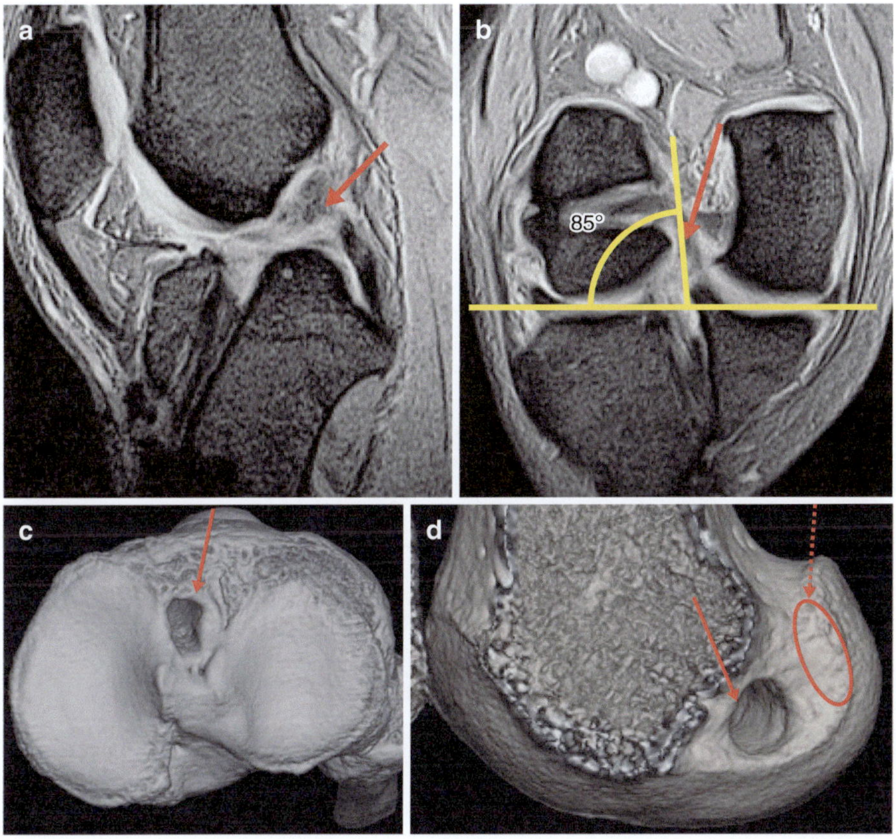

Fig. 10.3 An example of the poor ACL graft after non-anatomic reconstruction using the iliotibial band
(**a**) On the oblique sagittal MRI slice (T2*-weighted) in extension, a loose ligamentous shadow (arrow) is observed
(**b**) On the oblique coronal MRI slice (T2*-weighted) in extension, the ligamentous shadow (arrow) is verticalized and forms an angle of 85° to the tibial plateau
(**c**) 3-DCT image showing the tibial plateau, the tibial tunnel aperture (arrow) is almost at the anatomical attachment area
(**d**) 3-D CT image showing the lateral wall of the intercondylar notch. The round femoral tunnel (arrow) is located anteriorly and distally, far from the anatomic attachment area (dashed arrow, oval circle)

Table 10.1 Patient-based evaluation scoring system specific to treatment of ACL injury:JACL-25

Q01: Have you been bothered by the injured leg while standing in a train or bus?
☐ Not at all ☐ Slightly ☐ Moderately ☐ Considerably ☐ Extremely

Q02: Have you hesitated to walk briskly on a gravel or rough road?
☐ Not at all ☐ Slightly ☐ Moderately ☐ Considerably ☐ Extremely

Q03: Have you been afraid of walking on a slippery floor or road?
☐ Not at all ☐ Slightly ☐ Moderately ☐ Considerably ☐ Extremely

Q04: Have you been afraid of getting out of a car with your injured leg first?
☐ Not at all ☐ Slightly ☐ Moderately ☐ Considerably ☐ Extremely

Q05: Have you been afraid of jumping down from a 12-inch step, such as the steps of a bus?
☐ Not at all ☐ Slightly ☐ Moderately ☐ Considerably ☐ Extremely

Q06: Did you lack knee balance when performing a knee bend using only the injured leg?
☐ Not at all ☐ Slightly ☐ Moderately ☐ Considerably ☐ Extremely

Q07: Have you worried about running backward?
☐ Not at all ☐ Slightly ☐ Moderately ☐ Considerably ☐ Extremely

Q08: Have you been afraid of running down a slope?
☐ Not at all ☐ Slightly ☐ Moderately ☐ Considerably ☐ Extremely

Q09: Have you worried about running up the stairs two at a time?
☐ Not at all ☐ Slightly ☐ Moderately ☐ Considerably ☐ Extremely

Q10: Have you worried about running down the stairs?
☐ Not at all ☐ Slightly ☐ Moderately ☐ Considerably ☐ Extremely

Q11: Have you been bothered by the injured leg while sprinting?
☐ Not at all ☐ Slightly ☐ Moderately ☐ Considerably ☐ Extremely

Q12: Have you been afraid of jumping with the injured leg?
☐ Not at all ☐ Slightly ☐ Moderately ☐ Considerably ☐ Extremely

Q13: Have you been afraid of jumping and landing on the injured leg?
☐ Not at all ☐ Slightly ☐ Moderately ☐ Considerably ☐ Extremely

Q14: Have you been afraid of landing on both feet at one time?
☐ Not at all ☐ Slightly ☐ Moderately ☐ Considerably ☐ Extremely

Q15: Have you been afraid of performing a standing broad jump using only the injured leg?
☐ Not at all ☐ Slightly ☐ Moderately ☐ Considerably ☐ Extremely

Q16: Have you been afraid of a sudden change in direction (cutting) while on the injured leg?
☐ Not at all ☐ Slightly ☐ Moderately ☐ Considerably ☐ Extremely

Q17: Have you been afraid of hopping on the injured leg?
☐ Not at all ☐ Slightly ☐ Moderately ☐ Considerably ☐ Extremely

Q18: Did you protect the injured leg when stopping suddenly after sprinting?
☐ Not at all ☐ Slightly ☐ Moderately ☐ Considerably ☐ Extremely

Q19: Were you bothered by the injured leg during figure-of-eight running?
☐ Not at all ☐ Slightly ☐ Moderately ☐ Considerably ☐ Extremely

Q20: Have you worried about kicking a ball with the injured leg?
☐ Not at all ☐ Slightly ☐ Moderately ☐ Considerably ☐ Extremely

Q21: Have you worried about kicking a ball using the injured leg as a pivot?
☐ Not at all ☐ Slightly ☐ Moderately ☐ Considerably ☐ Extremely

Q22: Have you been afraid of someone bumping into your injured leg from the side?
☐ Not at all ☐ Slightly ☐ Moderately ☐ Considerably ☐ Extremely

Q23: Have you been afraid of performing a broad sideways jump with the injured leg?
☐ Not at all ☐ Slightly ☐ Moderately ☐ Considerably ☐ Extremely

Table 10.1 (continued)

Q24: Have you been afraid of taking a side step?				
☐ Not at all	☐ Slightly	☐ Moderately	☐ Considerably	☐ Extremely
Q25: Have you hesitated to slide with the injured leg?				
☐ Not at all	☐ Slightly	☐ Moderately	☐ Considerably	☐ Extremely

Adapted from Ref. [3]

This questionnaire asks about the condition of your injured knee during the past month. For each question, please check the box that best describes your condition

Please check that you have answered every question. Thank you for your cooperation

Each of the above items is scored from 0 to 4 and summed to give a score from 0 to 100; the closer to 0, the better.

POINTS

1. Most of the clinical reports in the literature are a list of "good/excellent" results. However, my own follow-up practices have rarely produced good results as described. This may be due to incomplete evaluation methods, the authors' bias, etc. Therefore, it is not easy to extract the truth from the literature. The only way to extract the truth is to see how far the logical statements are based on evidences and to cultivate the insight to detect illogical statements and erroneous conclusions, without being deluded by superficial and seductive conclusions. If you know the personalities of the authors of the papers, it may be easier to find the correct information/truths. For this reason, it may be helpful to meet the authors in person at academic conferences.
2. The indications for ACL reconstruction in patients with secondary mild to moderate osteoarthritis after a long period of time following ACL injury are controversial. Even if a good/successful ACL graft is obtained, the overall clinical results of the index knee may not be good. Honestly, the author was reluctant to recommend such patients undergo ACL reconstruction in the past. However, the author has changed the mind to perform it for such patients without treating chondral lesions, as these patients with stability-restores knees after the reconstruction were very much pleased about the stable knees despite of degenerative articular conditions. It seems that the functional disability caused by an unstable knee is much more annoying than that by the degenerative articular conditions. Patients must have felt instability at every step during walking.

References

1. Take Y, Shino K, et al. Early structural results after anatomic triple bundle anterior cruciate ligament reconstruction validated by tunnel location. Arthroscopy. 2018;34:2656–65.
2. Irrgang JJ, et al. Development and validation of the international knee documentation committee subjective knee form. Am J Sports Med. 2001;29:600–13.
3. Nagao M, et al. A novel patient-reported outcome measure for anterior cruciate ligament injury: evaluating the reliability, validity, and responsiveness of Japanese anterior cruciate ligament questionnaire 25. Knee Surg Sports Traumatol Arthrosc. 2016;24:2973–82.

Chapter 11
Revision ACL Reconstruction

Inappropriate/non-anatomic reconstructions are prevailing. ACL graft ruptures among patients who returned to sports after anatomic reconstruction are also common. Therefore, revision ACL reconstruction is often necessary. Occasionally, third or fourth reconstructions are performed.

11.1 Indications for Revision ACL Reconstruction

Regardless of the presence/continuity of the previously- reconstructed ligament, the patients suffering from instability due to the non-functioning ACL grafts, are candidates.

11.2 Problems with the Knees Requiring Revision ACL Reconstruction

In the knee with non-anatomical ACL graft, the articular cartilage and menisci are rarely in a healthy state because of the long-term persistence of instability. Therefore, various additional procedures such as meniscus repair as well as removal of the graft and fixation devices are necessary (Figs. 11.1, 11.2, and 11.3). Thus the revision should be performed by well-trained staff. To reduce the surgical invasion to

K. Shino, *Essence of Anterior Cruciate Ligament*,
https://doi.org/10.1007/978-981-99-6536-6_11

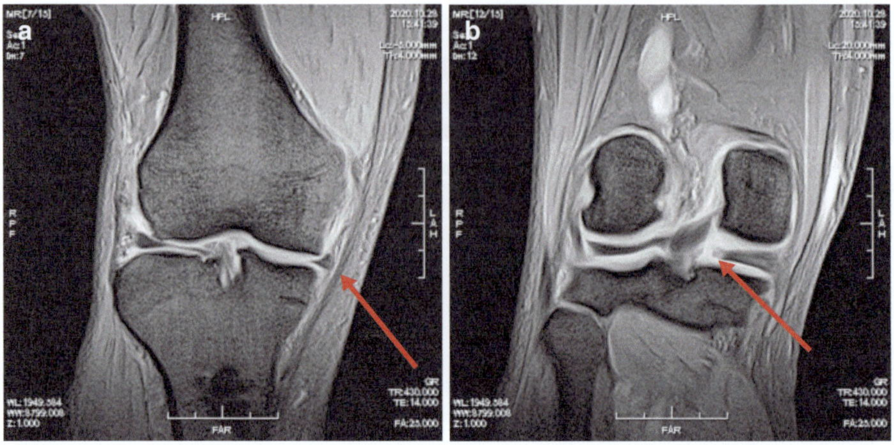

Fig. 11.1 Preoperative MRI findings of case TF (48-year-old male) (case presented in Fig. 10.3)
Twenty years ago, ACL reconstruction using the autologous iliotibial band in the right knee had been performed by the other surgeon. Since then, he managed to continue playing tennis with pain and swelling due to the unstable knee.
(**a**) The medial meniscus is extruded
(**b**) The posterior horn of the medial meniscus was torn
Therefore, we planned a pullout suture repair of the posterior horn through a small drill hole in the tibia

the affected limb, the use of a graft from the contralateral limb or allograft may be considered.

In cases with long-term persistence of unstable knees with poor / non-anatomical ACL graft, even if the stability is restored by the anatomical revision reconstruction, the patient may be suffering from arthritic symptoms due to damaged articular cartilage and loss of menisci.

On the other hand, in cases with ruptured anatomical graft, if the neglected period after rupture is not prolonged, the articular cartilage and menisci are often maintained in a healthy state. Thus they could restore the healthy knee free from arthritic symptoms due to damaged articular cartilage and loss of menisci, if the stability is restored by the anatomical revision reconstruction [1].

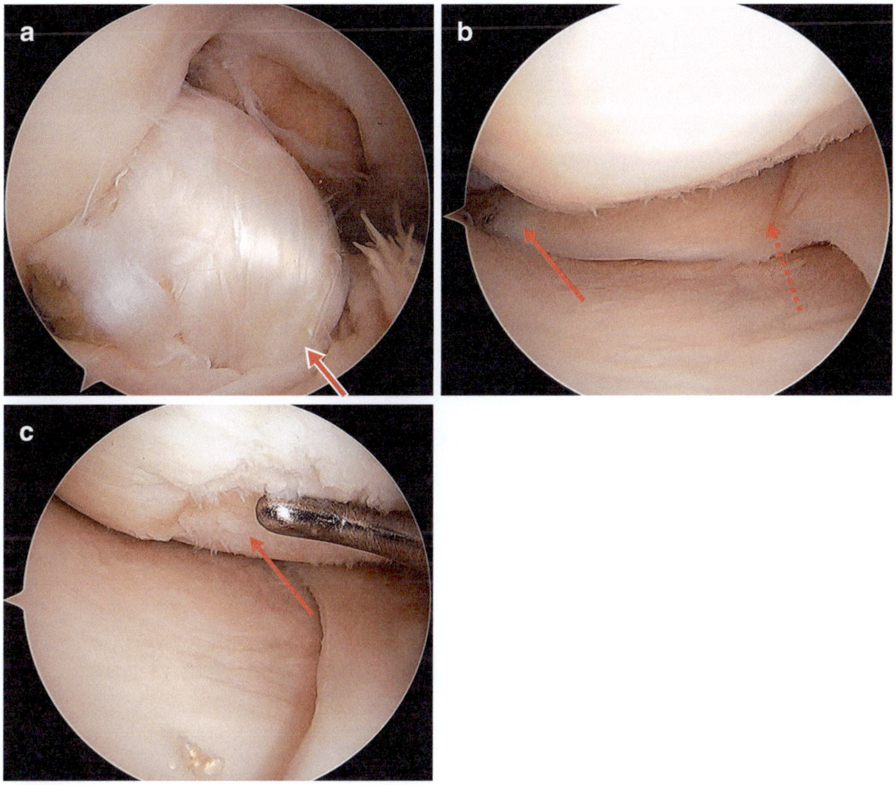

Fig. 11.2 Case TF: Arthroscopic findings during the revision surgery (viewed via the anterolateral portal)
(**a**) The primary ACL graft present in the intercondylar notch (arrow) was very lax and not functioning to stabilize the knee
(**b**) The posterior horn of the medial meniscus (arrow) was disrupted near its tibial attachment site, as shown as a wrinkle of the meniscus (dashed arrow)
(**c**) The articular cartilage of the medial condyle shows III° damage (arrow)
Thus, it is necessary to treat a variety of lesions during revision ACL reconstruction

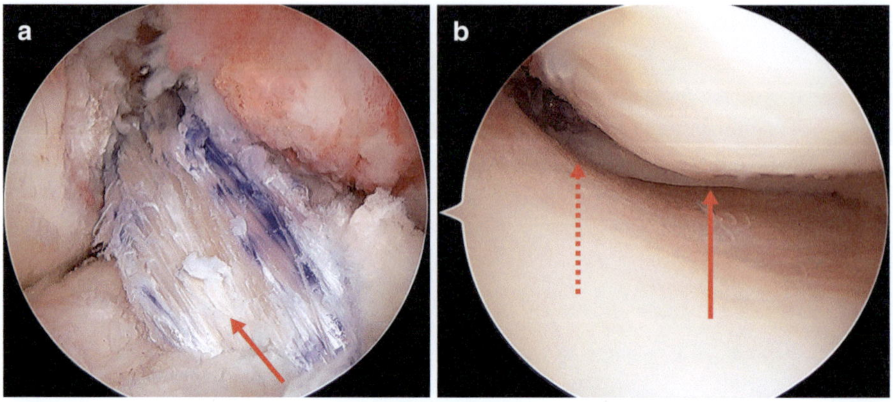

Fig. 11.3 Arthroscopic views of the revision ACL graft and the repaired meniscus of case TF (viewed via the anterolateral portal)
(**a**) The reconstructed ligament (arrow) with patellar tendon graft
(**b**) The repaired medial meniscus (arrow). The ruptured posterior horn was pulled into the bony tunnel created at the tibial attachment site, and secured with sutures
Note the wrinkle seen before disappeared

11.3 Fundamental Strategy for Revision Reconstruction

The anatomical tunnel should be accurately created for precisely-anatomical graft placement. The revision tunnels should not be influenced by the pre-existing tunnels at the previous surgery. If the tunnel position is compromised, the significance of revision will be questioned. Even if a partial tunnel overlapping occurs, anatomical graft placement should be pursued with ingenuity.

One-stage surgery should be planned. Please note that the patients who undergo revision surgery have already undergone several surgeries. Two-stage surgeries (e.g., first: bone graft into the previous tunnel; second: anatomical reconstruction) should be avoided to the greatest extent [2].

11.4 Preoperative Planning

If contractures or limited range of motion is present, removal of poorly-placed grafts and treatment of the damaged menisci is performed to normalize the range of motion as the first stage. This is the exceptional case for planning two-stage surgery.

The evaluation of the previously-reconstructed ACL is performed with not only MRI but also 3-D CT images, as described in “Imaging evaluation of the reconstructed ACL” of Chap. 10. The latter is very helpful to evaluate the position, shape, and enlargement of the bony tunnel apertures. This makes it easy to plan creating anatomical ACL reconstruction tunnels (Fig. 11.4).

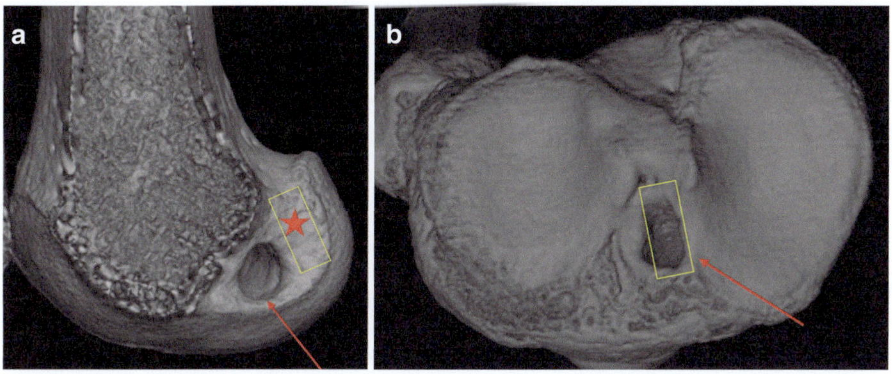

Fig. 11.4 3-D CT Image of the right knee for preoperative planning the anatomical rectangular tunnel ACL reconstruction (case TF)

(**a**) The lateral wall of the intercondylar notch. The femoral tunnel aperture (arrow), is not present within the anatomic attachment area. A new rectangular tunnel behind it was planned to create (asterisk)

(**b**) The tibial plateau. The tibial tunnel aperture (arrow) is almost entirely within the anatomic attachment area. So it is planned to enlarge anteroposteriorly

11.5 Graft Options

1. Graft harvest from the opposite side

   Currently, as allogeneic tendons from cadavers are not available in Japan, autogenous tendons are generally used. The use of a tendon graft from the contralateral side is advantageous in shortening the operation time and minimizing surgical invasion on the affected side. This makes it easier to perform postoperative rehabilitation.

2. Tendons with bone plugs or those without bone plugs

   As described in Chap. 7, considering the stress concentration on the graft at the femoral tunnel aperture, a tendon graft with bone plug(s) is preferred.

   The bone-patellar tendon-bone graft is preferred to the quadriceps tendon-bone in terms of muscle recovery.

3. The author's autograft selection sequence for revision

   (1) The ipsilateral patellar tendon with bone plugs
   (2) Contralateral patellar tendon with bone plugs (if ipsilateral patellar tendon has already been harvested)
   (3) Ipsilateral quadriceps tendon (with a bone plug) (5~6 mm thick)
   (4) Contralateral quadriceps tendon (with a bone plug) (5~6 mm thick)
   (5) Ipsilateral or contralateral hamstring tendon

11.6 Surgical Procedure

Anatomical rectangular tunnel ACL reconstruction with a bone-patellar tendon-bone graft or the quadriceps tendon-bone graft is the fundamental procedure for the author (Fig. 11.5) [1–4]. This is due to the following reasons:

1. Bone plugs provide early integration of the graft to the tunnel wall [5].
2. By aligning the bone-tendon junction with the femoral tunnel aperture, the excessive stress concentration caused by the graft-bending angle (approximately 80°) is accommodated.
3. The smaller cross sectional area of the rectangular tunnel allows for easier avoidance of tunnel overlapping. For creating tunnel for a graft with 10-mm diameter, the rectangular tunnel has a cross section of 50 mm^2 , while the conventional round tunnel has that of 79 mm^2.

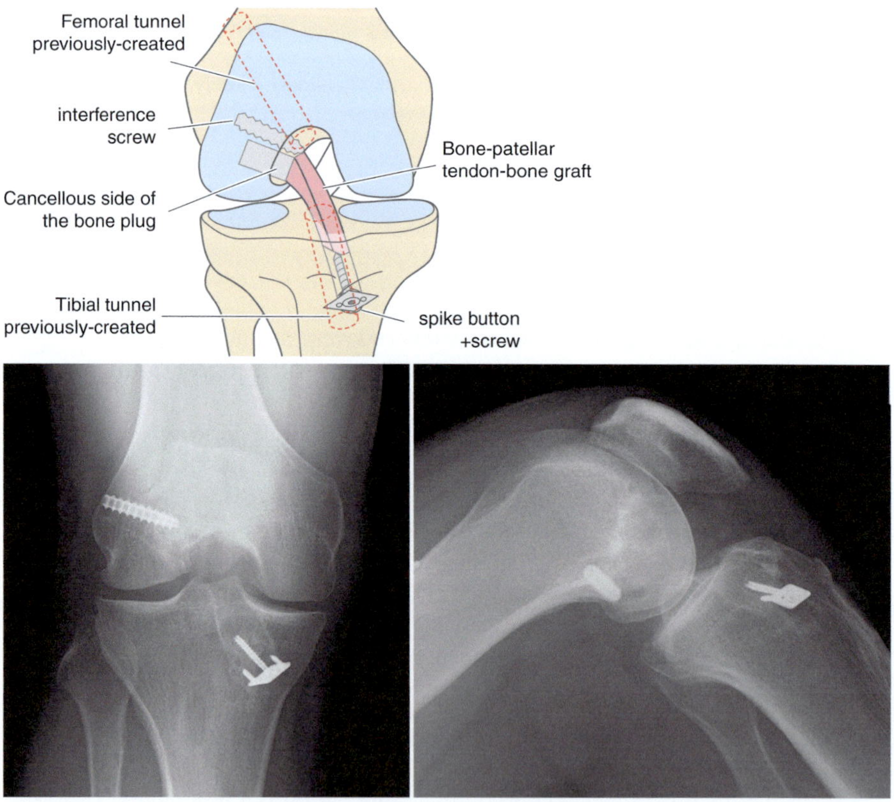

Fig. 11.5 Schema of the anatomical rectangular tunnel revision ACL reconstruction using a bone-patellar tendon-bone graft
Thanks to the smaller cross-sectional area of the tunnels in this technique, it is easier to create the anatomical tunnels without tunnel overlapping
(Based on Refs. [1–4])

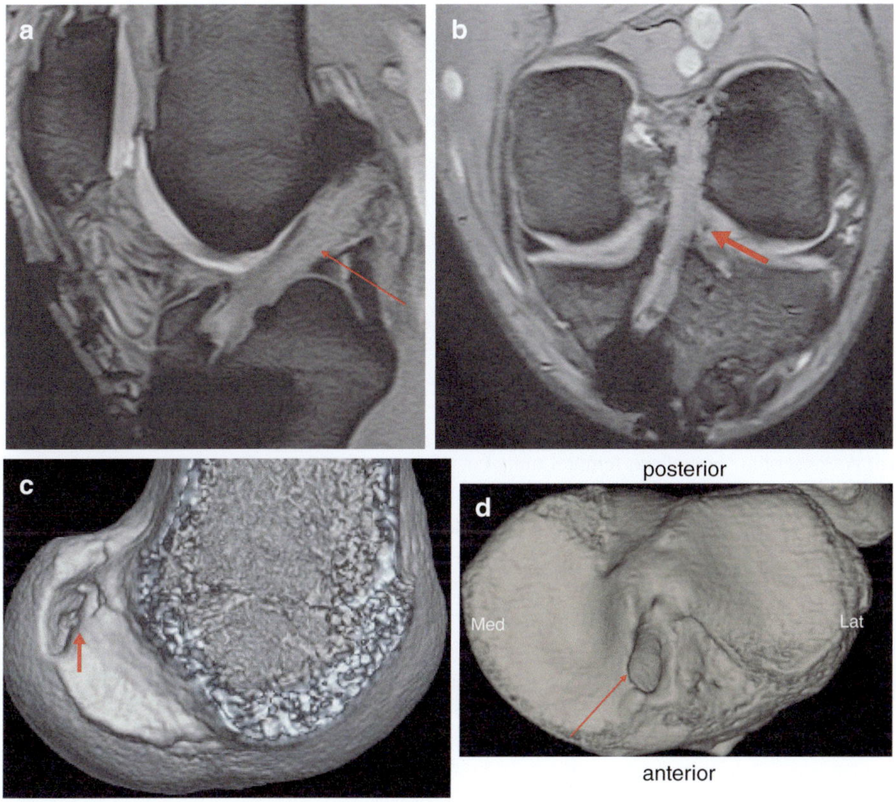

Fig. 11.6 MRIs (T2*-weighted) and 3-D CT images at 9 months after left revision anatomical rectangular tunnel ACL reconstruction using the contralateral/right patellar tendon graft due to the primary anatomical graft rupture
Both sagittal (**a**) and coronal (**b**) sections of MRI show good reconstructed ligaments (arrows). The signal intensity is still higher than normal. (**c**) Femoral tunnel aperture (arrow), (**d**) Tibial tunnel aperture (arrow) are shown on the 3-D CT. These tunnels were the same tunnels as those in the primary surgery

1. In case the previous surgery is performed anatomically

Since the same tunnel can be used, the procedure should be performed as in the previous reconstruction (Fig. 11.6). If the hamstring tendon was used previously, the tunnel may be enlarged in the anterior/distal direction. For such cases, an interference screw may need to be placed anteriorly to the bone plug in the femoral tunnel.

2. If the primary surgery is non-anatomically performed

As the non-anatomical tunnels cannot be used, new anatomical tunnels should be created without tunnel overlapping [1, 3].

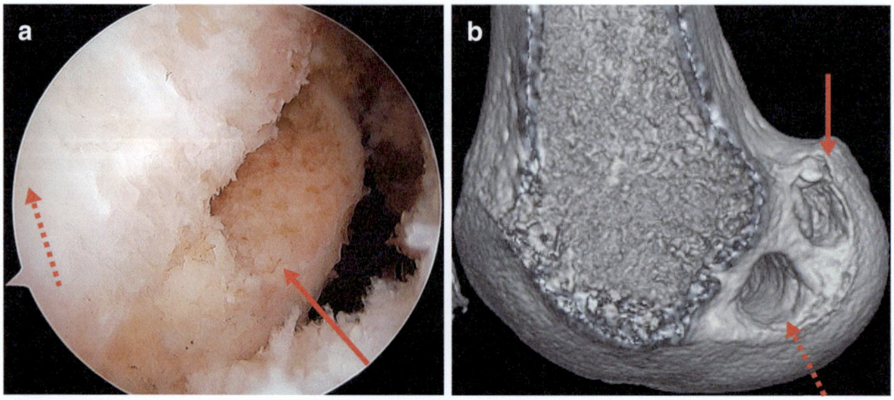

Fig. 11.7 Femoral tunnel aperture in case TF
(**a**) Arthroscopic view of the femoral tunnel at the time of revision. The new rectangular tunnel (arrow) does not interfere with the previous tunnel (dashed arrow)
(**b**) 3-D CT image at 3 weeks. The new rectangular tunnel (arrow) does not interfere with the previous tunnel (dashed arrow)

(1) Femoral side

It is almost possible to create a bony tunnel posterior to the previous tunnel avoiding tunnel overlapping (Fig. 11.7). If the minimum distance between the two tunnel apertures is less than 5 mm, the primary reconstruction tunnel can be filled using an interference screw [2] (Fig. 11.8).

In cases where the anatomical tunnel cannot be created, the graft is guided out of the joint via the over-the-top of the lateral condyle and fixed to the posterior/lateral femoral cortex above the lateral condyle [3] (Fig. 11.9).

(2) Tibial side

In most cases, it is possible to reuse the previous tunnel (Fig. 11.10).

If the tunnel is placed posteriorly, it is often possible to create the new one anterior to the previous tunnel without tunnel overlapping (Fig. 11.11a). However, partial tunnel overlapping is not uncommon. If the overlap is less than 4 mm, the bone plug is held in the center of the tunnel by pulling out the bone plug along the tunnel axis, and fixed with the BTF system (see Fig. 7.13). This allows anatomic placement of the graft (Fig. 11.11b). If the overlapping is greater than 5 mm, the one plug is eccentrically/anatomically positioned, and the void is filled with bone graft (Fig. 11.11c).

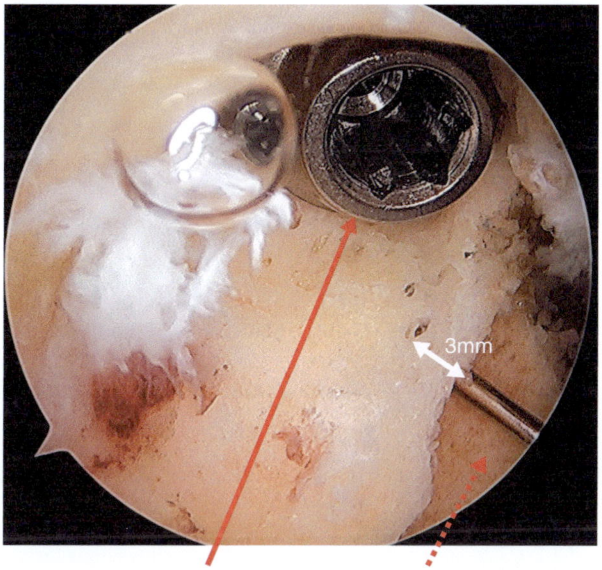

Fig. 11.8 Reinforcement of the anterior wall of the new femoral tunnel using an interferential screw (right knee, viewed through the anteromedial portal)
When the septal wall of the two femoral tunnel apertures, old and new, was less than 5 mm, the previous non-anatomic tunnel was filled with an 8–9 mm diameter interference screw to reinforce the anterior wall of the new tunnel

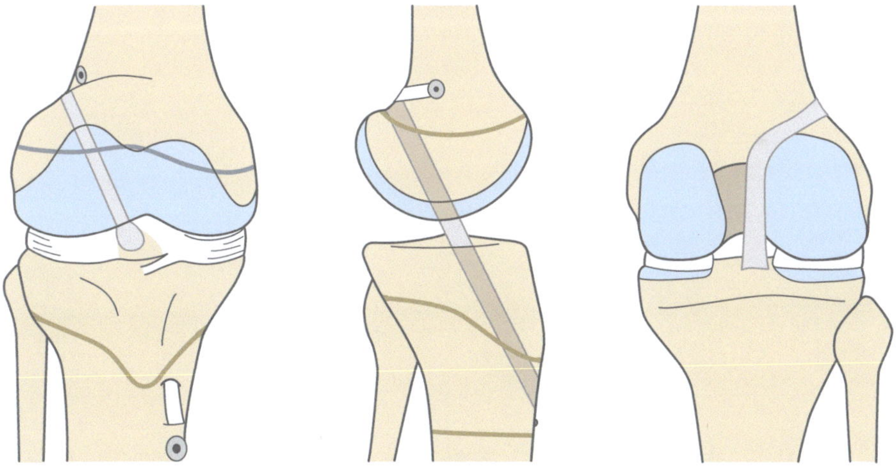

Fig. 11.9 Over-the-top ACL (revision) reconstruction/right knee
In cases where the anatomical tunnel cannot be created because a large bony defect has developed at the femoral attachment site due to multiple reconstructions, the graft is guided out of the joint via the over-the-top of the lateral condyle and fixed to the posterior/lateral femoral cortex above the lateral condyle. (Modified from Ref. [3])

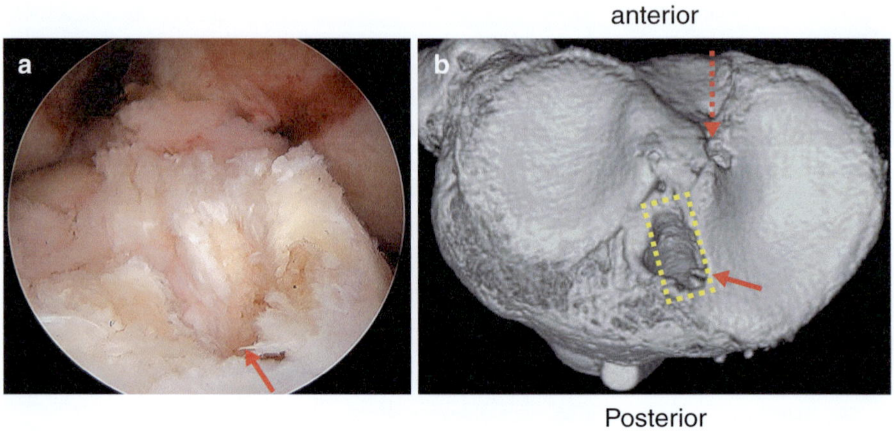

Fig. 11.10 Rectangular tibial tunnel aperture during the revision reconstruction for case TF (right knee)
(**a**) Arthroscopic view from the anteromedial portal (arrow) (view through the anteromedial portal)
(**b**) 3-D CT image on the tibial plateau. The previous tunnel is expanded anteriorly and posteriorly to create the new rectangular tunnel (arrow). A small round tunnel to pullout the posterior horn of the medial meniscus can be seen (dashed arrow)

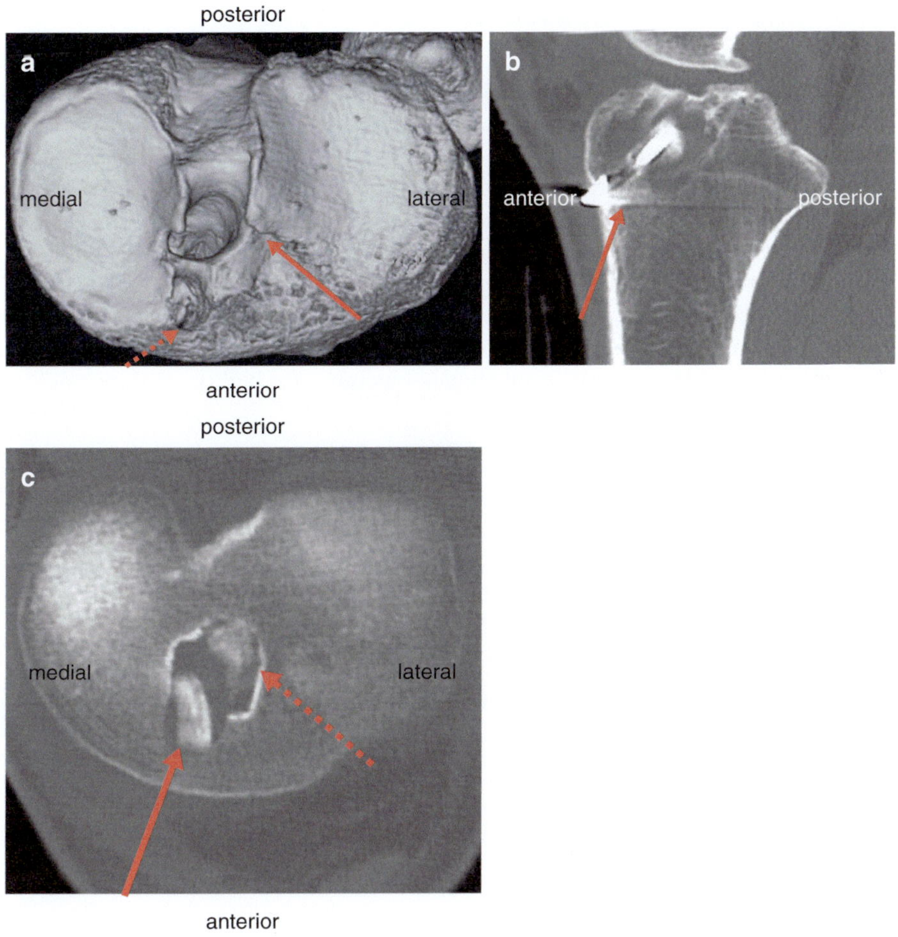

Fig. 11.11 Creation of tibial tunnel in revision reconstruction (fourth reconstruction, left knee)
(**a**) The new tunnel (dashed arrow) anterior to the previous posteriorly-placed tunnel (arrow) was created without tunnel overlapping (third reconstruction, left knee). While the tunnel protruded several millimeters anteriorly from the attachment area, there was no graft impingement to the intercondylar notch even in full extension intraoperatively
(**b**) An example of partial tunnel overlapping of 4 mm or less. The bone plug was fixed using BTF system/Spike button, Intra-bone plug screw (Fig. 7.13) to hold the bone fragments (arrows) in the center of the tunnel, allowing anatomical graft placement (4th reconstruction, left knee)
(**c**) An example of tunnel overlapping of 5 mm or greater. The bone plug (arrow) was eccentrically positioned to the anatomical location. The void was filled with bone graft (dashed arrow) (4th reconstruction, left knee)

11.7 Postoperative Rehabilitation, Prognosis

- Depending on the gap between the graft and the tunnel wall, the initiation and progression of range-of-motion exercise may be delayed.
- Patients with a long history of unstable knees who had undergone non-anatomic reconstruction often have deteriorated articular cartilage and menisci (Fig. 11.2). Even for such cases, the properly-performed revision anatomical reconstruction is effective to restore stability without additional extra-articular tenodesis.
- When sports activity level is elevated after restoring knee stability by successful revision reconstruction, arthritic symptoms due to deteriorated articular cartilage and menisci during strenuous activity may develop. Good reconstructed ligaments do not always lead to good clinical outcomes for highly active people.

References

1. Shino K, et al. Surgical technique: revision ACL reconstruction with a rectangular tunnel technique. Clin Orthop Relat Res. 2012;470:843–52.
2. Shino K, et al. One-stage revision anatomic anterior cruciate ligament reconstruction with rectangular tunnel technique. Asia Pac J Sports Med Arthrosc Rehabil Technol. 2015;2:43–8.
3. Shino K, et al. How to handle a poorly placed femoral tunnel. In: Marx RG, editor. Revision ACL reconstruction: indications and technique. New York: Springer; 2014. p. 87–96.
4. Shino K, et al. Rectangular tunnel double-bundle anterior cruciate ligament reconstruction with bone-patellar tendon-bone graft to mimic natural fiber arrangement. Arhroscopy. 2008;24:1178–83.
5. Suzuki T, Shino K, et al. Early integration of a bone plug in the femoral tunnel in rectangular reconstruction with a bone-patellar tendon-bone graft: a prospective computed tomography analysis. Knee Surg Sports Traumatol Arthrosc. 2011;19(Suppl 1):S29–35.

Chapter 12
Keys for Success

12.1 Timing of Surgery

The effect of delay for several weeks in surgical timing on the healing of the reconstructed ligament is negligible. However, cases with unstable knees left untreated for a prolonged period of time have damage to the menisci and articular cartilage (Figs. 11.1 and 11.2). Therefore, ACL-injured knees should not be left unstable for a long period to preserve the structures inside the joint.

1. Acute Phase

In general, surgery should be performed after the range of motion is normalized by aspiration of hemarthrosis and rehabilitation (Fig. 12.1).

In case with a severe traumatic injury such as ACL injury, it is known that the healing response is activated for several weeks after the injury. Therefore, if meniscus repair is performed during this period, the healing rate will be higher (Figs. 12.2, 12.3, and 12.4).

Patients with combined medial collateral ligament injury often show poor range of motion, probably due to inflammation/pain caused by the vigorous healing process of the medial aspect. ACL reconstruction in such a situation is likely to lead to joint contracture/arthrofibrosis (Fig. 12.5). When early reconstruction is unavoidable (e.g., associated with complex meniscus injury), the utmost attention should be paid to postoperative rehabilitation to prevent/arthrofibrosis.

If there is no combined complex/serious meniscus injury, conservative treatment of the injured medial collateral ligament is applied using a functional brace to restrict external tibial rotation or abduction. ACL reconstruction is performed after normalization of range of motion and healing of the medial collateral ligament.

K. Shino, *Essence of Anterior Cruciate Ligament*,
https://doi.org/10.1007/978-981-99-6536-6_12

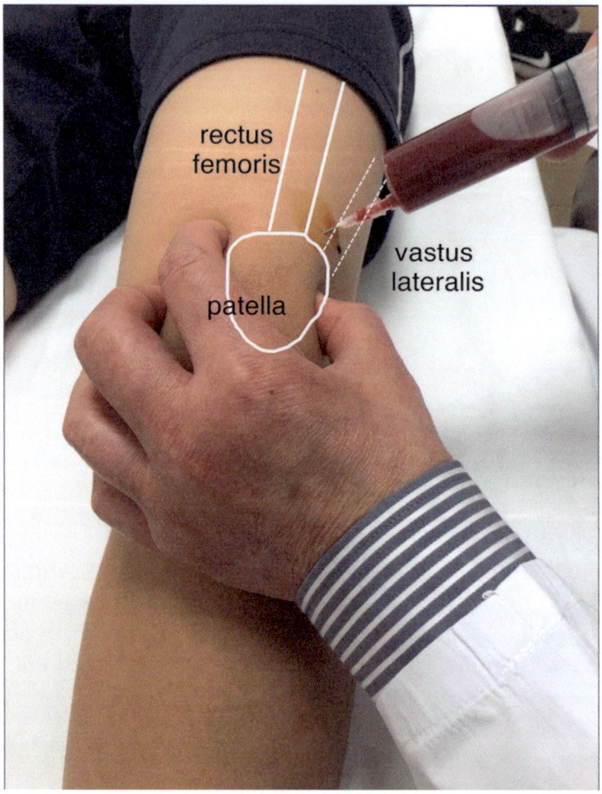

Fig. 12.1 Aspiration of hemarthrosis.
The most reliable site for arthrocentesis is between the rectus femoris and vastus lateralis tendons. Perpendicular puncture to this site ensures intra-articular entry

In cases with full thickness articular cartilage injury, mesenchymal stem cells from bone marrow may lead to spontaneous healing. Therefore, it is not necessary to perform a cartilage transplantation simultaneously (Figs. 12.6, 12.7, 12.8, and 12.9). If spontaneous healing cannot be achieved after several months (during the rehabilitation period after ACL reconstruction), further treatment including cartilage transplantation may be considered.

2. Chronic phase

Patients with long history of unstable knees often present with locking symptom due to torn medial meniscus (Fig. 4.1). If the locking is continued for a long period of time, contracture is likely to remain even after the locking is released by treatment of the injured meniscus. If so, rehabilitation for several weeks is performed to resolve the joint contracture prior to subsequent ACL reconstruction.

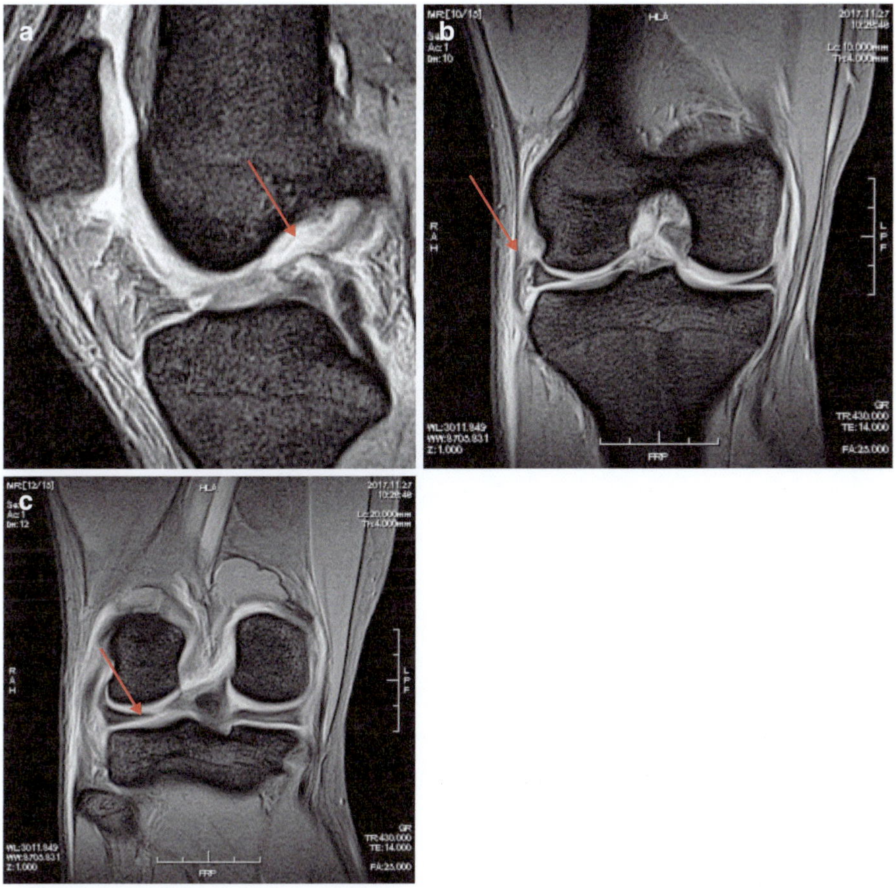

Fig. 12.2 MRI (T2*-w) image of the right knee of case MT (21 years old, male, collegiate rugby football player). Injured by abduction stress
(**a**) Injured ACL; (**b**) Extrusion of the lateral meniscus; (**c**) Disruption of the posterior segment of the lateral meniscus. Early surgery was planned

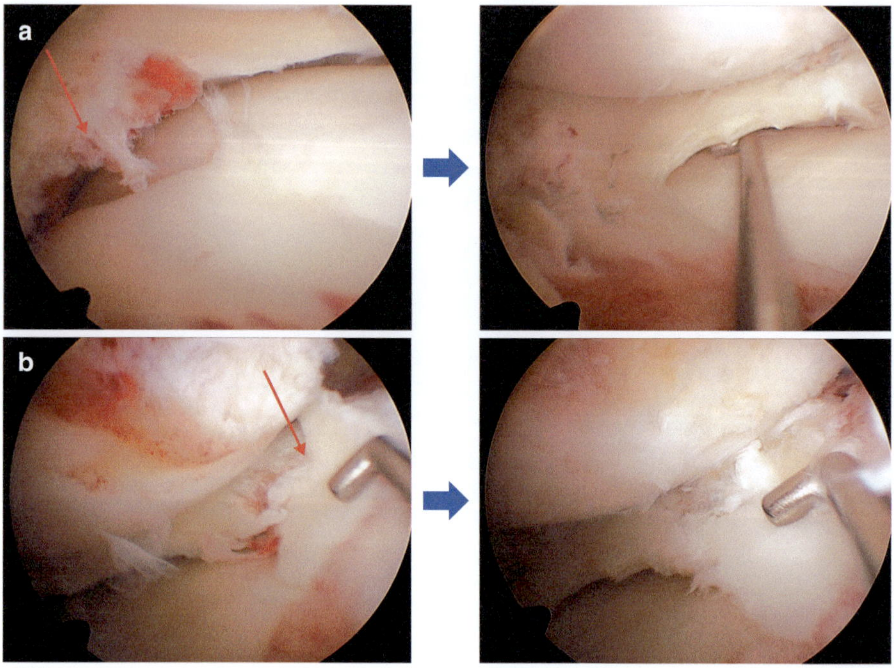

Fig. 12.3 Intraoperative findings of case MT
(**a**) L-shaped tear of the middle segment of the lateral meniscus (arrow) was sutured with inside-out technique
(**b**) Transverse tear of the posterior segment of the lateral meniscus (arrow) was sutured with all-inside suture technique with a suture passer

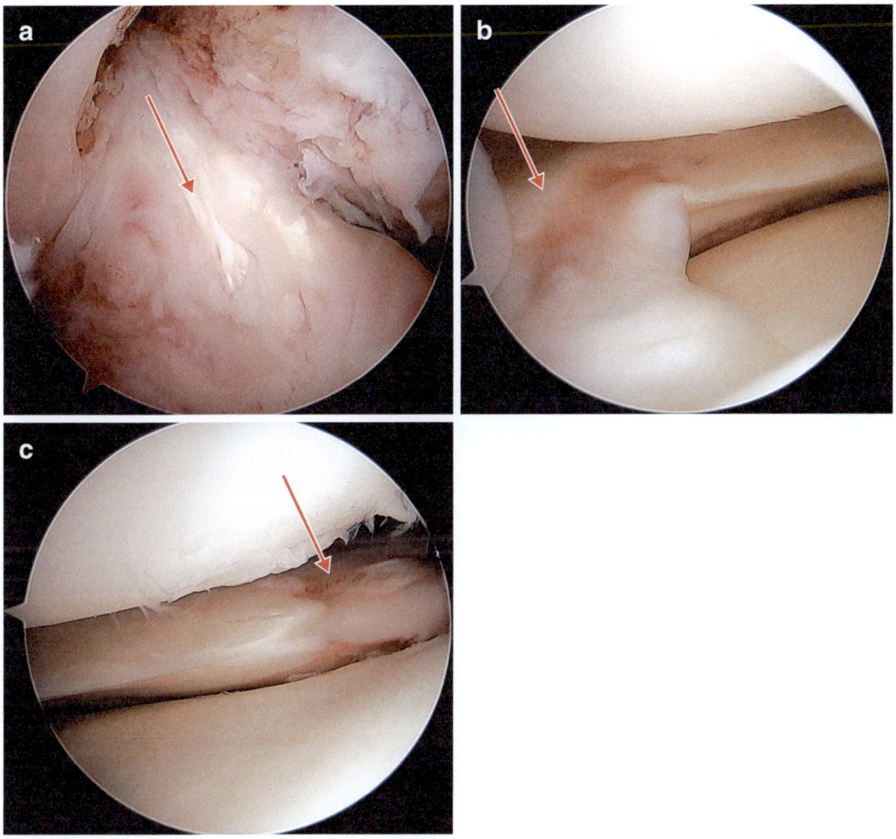

Fig. 12.4 Second-look arthroscopy on case MT at 4 months postoperatively
(**a**) Reconstructed ACL with the patellar tendon graft (arrow)
(**b**) Healed middle segment of the lateral meniscus (arrow)
(**c**) Healed posterior segment of the lateral meniscus (arrow)
The patient fully returned to competitive rugby football at 9 months after the surgery, and successfully completed his sports activities as a captain of the team until his graduation in the following year

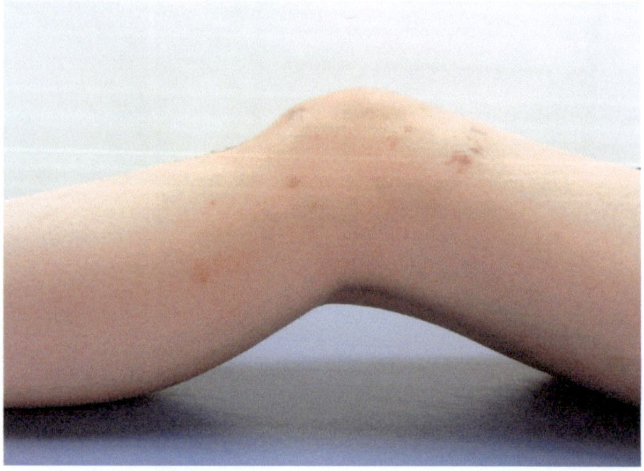

Fig 12.5 Arthrofibrosis Postoperatively, bleeding from the bone marrow is inevitable, and immobilization is necessary. Thus there is a risk of arthrofibrosis/joint contracture. Needless to say, prevention is important

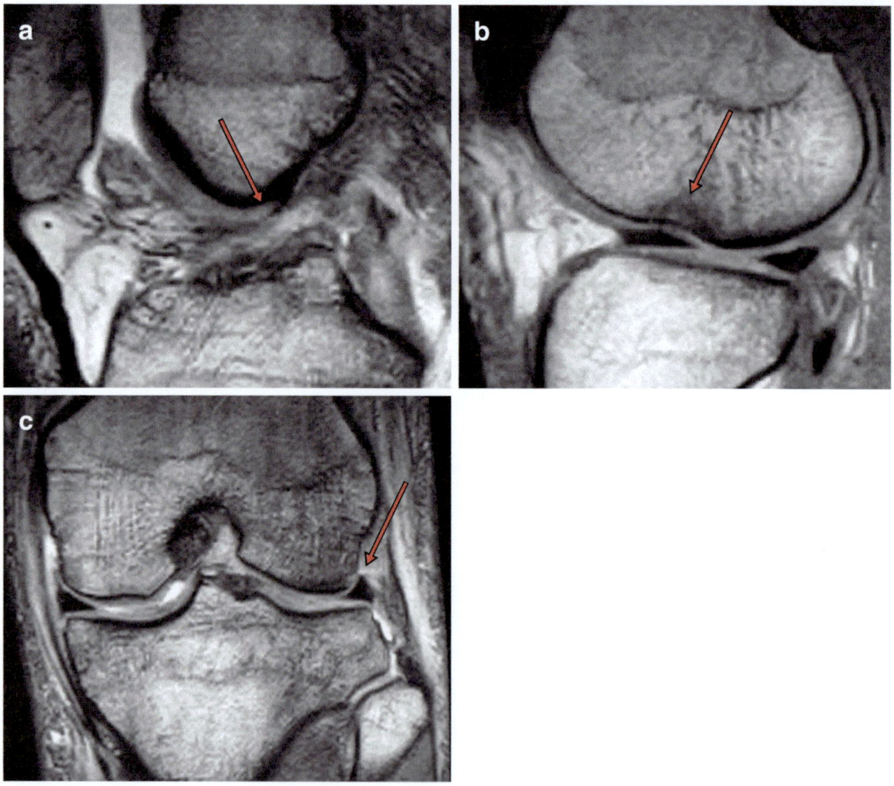

Fig. 12.6 MRI (T2*-weighted) of the left knee in case TY (19-year-old male American football player) injured with abduction stress
(**a**) ACL injury (arrow)
(**b**) Bone bruise of the lateral femoral condyle (arrow)
(**c**) extrusion of the lateral meniscus (arrow)

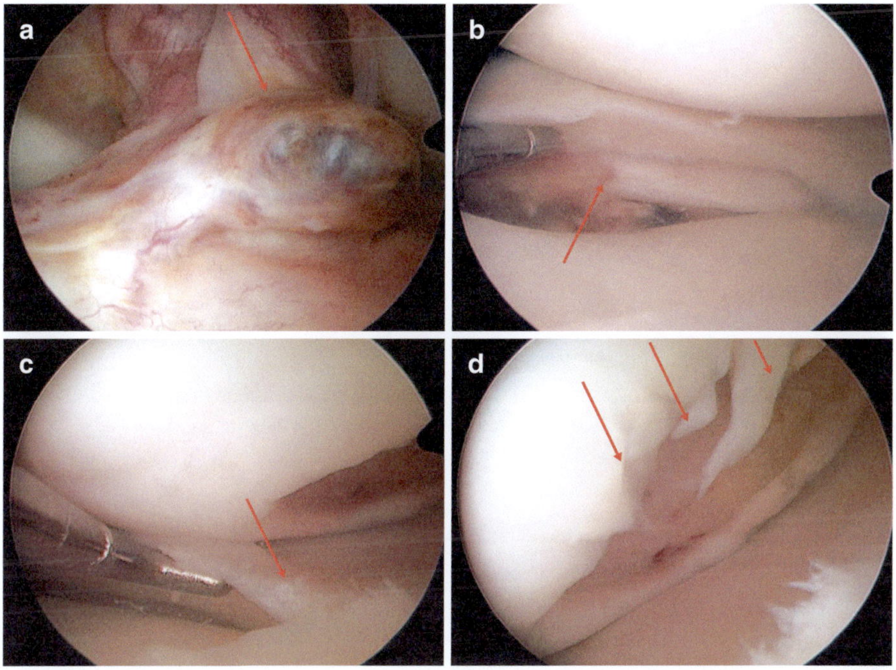

Fig. 12.7 Intraoperative findings in case TY
(**a**) Injured ACL (arrow)
(**b**) Longitudinal tear of the posterior segment of the lateral meniscus (arrow)
(**c**) Radial split tear of the middle segment of the lateral meniscus (arrow)
(**d**) Cartilage defect of the lateral femoral condyle (arrows)

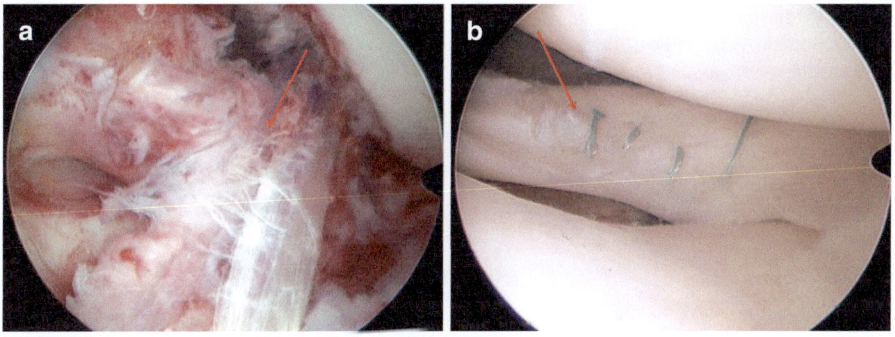

Fig. 12.8 Surgery for case TY
(**a**) Anatomic triple-bundle ACL reconstruction with hamstring tendon graft (arrow), (**b**) Partial resection of the transverse tear of the middle segment followed by suturing the posterior segment of the lateral meniscus with inside-out technique (arrow), leaving the cartilage defect untreated (in the expectation of spontaneous healing, since bleeding from the bone marrow was seen)

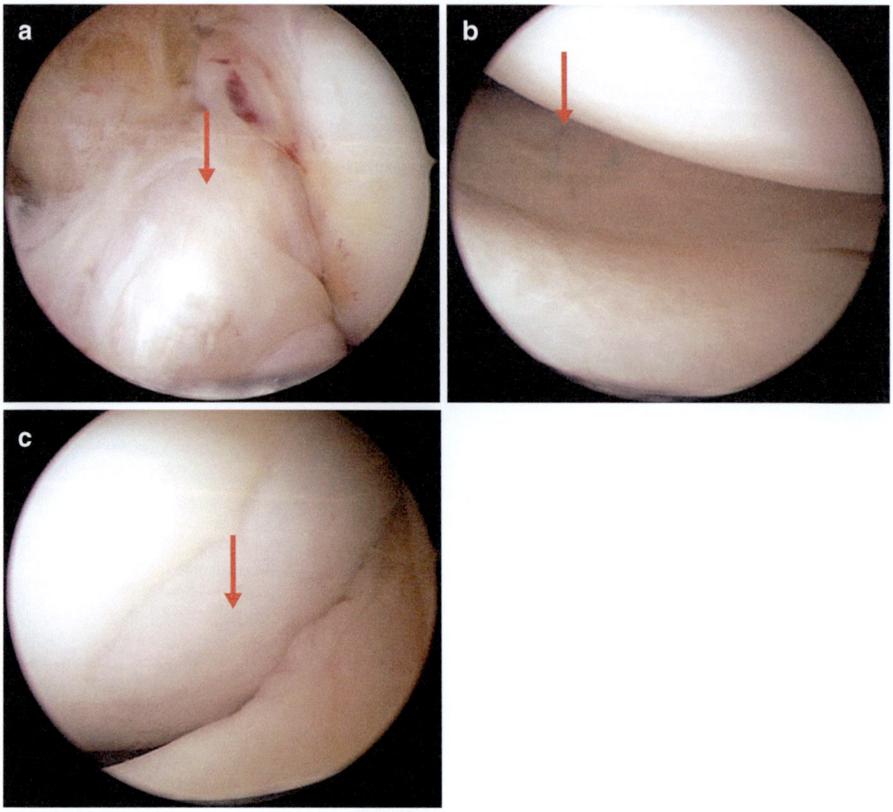

Fig. 12.9 Second-look arthroscopy on case TY at 3 months postoperatively
(**a**) Reconstructed ACL with hamstring tendon (arrow)
(**b**) Healed posterior segment of the lateral meniscus (arrow)
(**c**) Spontaneously healed cartilage lesion of the lateral femoral condyle. The defect was filled with hard cartilage-like tissue (arrow)

12.2 Intraoperative Complications

The authors perform surgery under a pneumatic tourniquet by holding the thigh horizontally with a leg holder and allowing the calf to droop (Fig. 12.10). In this case, not only the pneumatic tourniquet but also the posterior aspect of the thigh is compressed by gravity. Prolonged surgery (about 3 h) in muscular patients may cause crush syndrome due to damage to the thigh muscles. In those patients, use of the pneumatic tourniquet should be limited to about 2 h.

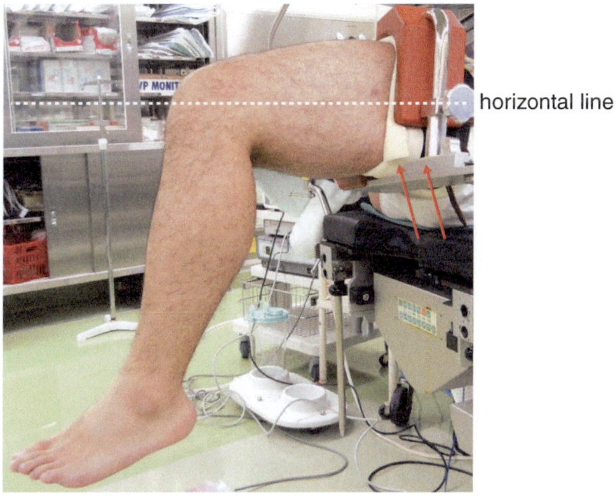

Fig. 12.10 Positioning for arthroscopic surgery
Using a leg holder, the thigh is kept horizontally with the calf hung down, and a pneumatic tourniquet is applied. This setting makes it easier to perform every kind of arthroscopic surgery. However, If surgery is prolonged in this positioning, crush syndrome may occur as the posterior aspect of the thigh is kept compressed by gravity (arrow)

12.3 Postoperative Complications

1. Infection

Infection often becomes apparent 1 to several weeks after surgery. It goes without saying that thorough sterilization of surgical instruments is important for prevention. Early detection → MRI to confirm the lesion → early treatment (intra-articular debridement, drainage, chemotherapy) leads to cure. Removal of the graft is rarely required (Fig. 12.11). However, in rare cases, when the graft itself is evaluated to be infected based on MRI and arthroscopic findings, it must be removed to subside the infection.

2. Joint contracture/Arthrofibrosis

Postoperatively, bleeding from the bone marrow is inevitable, and immobilization is necessary. Thus there is a risk of arthrofibrosis/joint contracture. Needless to say, prevention is important (Fig. 12.5). Thus there is a risk of arthrofibrosis/joint contracture to be prevented.

Risk factors include acute surgery, concomitant medial collateral ligament injury, meniscus repair, and female patients.

As the response or tolerance to pain varies from person to person, improvement in range of motion one week after removal of immobilization must be carefully checked for further intervention in rehabilitation. There is no quick remedy for arthrofibrosis, so rehabilitation must be continued steadily. As the functional position of the knee joint is full extension, the knee must not lose even 1 degree of extension.

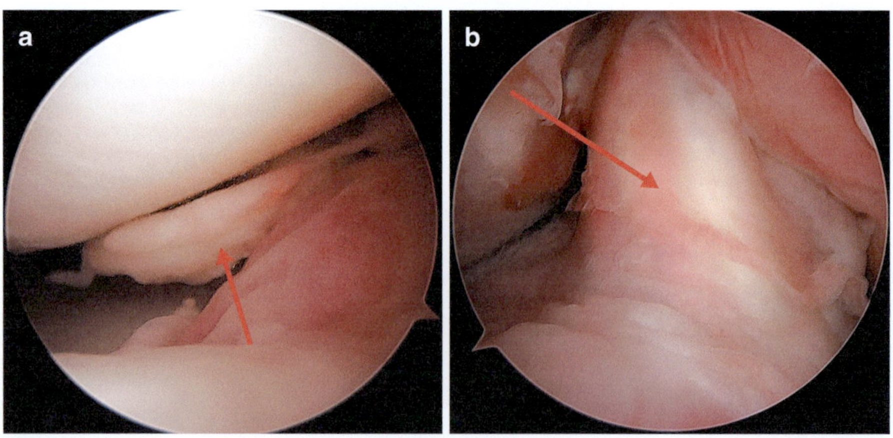

Fig. 12.11 Second-look arthroscopy for a case in which symptoms of infection became apparent 3 weeks after the anatomical triple-bundle ACL reconstruction using hamstring tendon graft (right knee)
(**a**) Debris is floating in the joint (arrow)
(**b**) Reconstructed ACL using hamstring muscle tendon (arrow) is completely covered with synovial membrane

POINT

It goes without saying that early detection and treatment are the key to prevent arthrofibrosis/joint contracture. In a few months, some cases may develop excessive fibrocartilaginous tissue formation in the intercondylar notch, causing pain on extension (Fig. 12.12). For such cases, it is necessary to remove the scar tissue in the intercondylar notch.

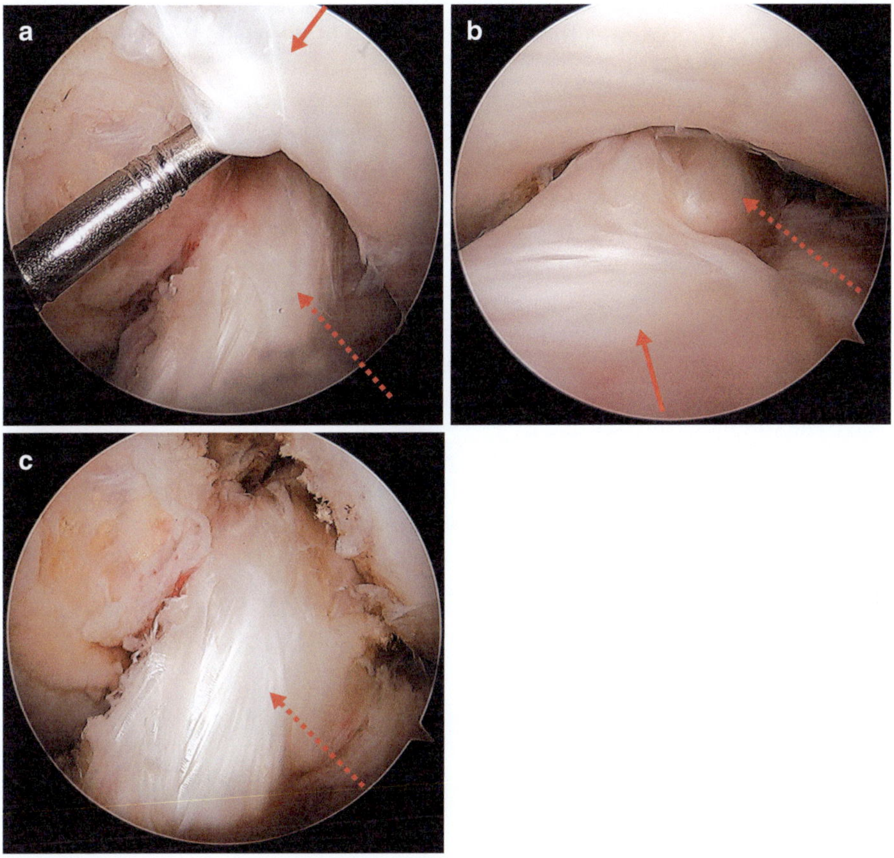

Fig. 12.12 Example of a 21-year-old male with mild limitation in extension after ACL reconstruction.
After the anatomical rectangular tunnel ACL reconstruction with patellar tendon graft on the right knee, the patient was suffering from mild limitation in extension and pain on extension. Twelve months after surgery, he underwent a subsquent surgery to resolve the issues as well as to remove the hardware used for graft fixation
(**a**) Fibrocartilaginous tissue growth in the intercondylar notch (arrow)
(**b**) Fibrocartilaginous tissue growth anterior to the ACL graft (arrow)
(**c**) Healthy thick ACL graft after removal of the fibrocartilaginous tissues with RF device (dashed arrow)
The patient returned to judo without pain in extension

12.4 Secondary ACL Injury

Even if patients return to their pre-injury sports with good restored stability, more than 10% of patients sustain injury to an ipsilateral ACL graft or contralateral ACL. The younger the patient, the higher the re-injury rate. Therefore, re-injury prevention is a major issue [1].

There is no perfect prevention measures. However, an approach to risk factors must not be abandoned. The followings are the main points.

1. Strengthen the postural control muscles (trunk and abdominal muscles) with the aim of quickly recovering from the posterior tilt of the trunk, which is a dangerous position to ACL injury [2].
2. Select shoes or spikes appropriate for the respective court or ground. *Once a right handed tennis player sustained an ACL injury of the pivot/right leg while hitting a backhand shot on a hard court wearing shoes for clay court.*
3. Adjust safety equipment such as ski bindings properly.
4. Thoroughly educate the patients on the risk of (re)injury from (1) to (3) above.

POINT

Various ACL injury prevention programs (consisting of three types of exercises : strength, balance, and jump) have been proposed [3].

The strength exercises mainly focus to strengthen the quadriceps and hamstring muscles; the balance exercises aim to improve sense of balance using a balance disc; the jump exercises are to optimize the alignment of the lower limb during landing. However, the effectiveness of the above-mentioned exercises has not been thoroughly proven, while it is also known that education on injury risk alone has a preventive effect.

References

1. Mae T, Shino K, et al. Risk factors for ipsilateral graft rupture or contralateral anterior cruciate ligament tear after anatomic double-bundle reconstruction. Asia Pac J Sports Med Arthrosc Rehabil Technol. 2014;1:90–5.
2. Koyanagi M, Shino K, et al. Effects of changes in skiing posture on the kinetics of the knee joint. Knee Surg Sports Traumatol Arthrosc. 2006;14:88–93.
3. Myklebust G, et al. Prevention of anterior cruciate ligament injuries in female team handball players: a prospective intervention study over three seasons. Clin J Sport Med. 2003;13:71–8.

Correction to: Essence of Anterior Cruciate Ligament: Scientific Evidence and Clinical Practice of ACL

Correction to: K. Shino, *Essence of Anterior Cruciate Ligament*: Scientific Evidence and Clinical Practice of ACL, https://doi.org/10.1007/978-981-99-6536-6

The book was inadvertently published with incorrect copyright information. The copyright year was missed on the book copyright page for the Japanese language edition. The copyright holder has been updated in the book.

Based on a translation from the Japanese language edition:

ACL no Essence Hiza Zenjujijintai no Evidence to Rinsho by Konsei Shino Copyright © Bunkodo Co., Ltd. 2022

The updated version of the book can be found at https://doi.org/10.1007/978-981-99-6536-6

K. Shino, *Essence of Anterior Cruciate Ligament*,
https://doi.org/10.1007/978-981-99-6536-6_13

Afterword/Acknowledgments

I was motivated to write this book because I felt that the hypothesis regarding anatomical ACL reconstruction, which I have been studying for the past 20 years, has been supported by scientific evidences, and that a highly reproducible and accurate surgery can be realized clinically, and that a treatment system including postoperative rehabilitation has been established.

This book is the fruit of the efforts of my fellow physiotherapists, my younger colleagues at Osaka University Medical School, my former fellows at Sapporo Medical University who continued to work with me after their training period in Osaka, and Dr. Fujie and his laboratory staffs at Tokyo Metropolitan University. I would like to take this opportunity to express my deepest gratitude to all of them. I would also like to express my deepest gratitude to Mr. Haruhiko Nakamura of Bunkodo in Japanese Edition and Ms. Saki Kasai and Makie Kambara of Springer in English Edition for their great effort and encouragement in publishing this book.

December 2023
Konsei Shino, MD, PhD

K. Shino, *Essence of Anterior Cruciate Ligament*,
https://doi.org/10.1007/978-981-99-6536-6

MIX
Papier aus verantwortungsvollen Quellen
Paper from responsible sources
FSC® C105338

If you have any concerns about our products,
you can contact us on
ProductSafety@springernature.com

In case Publisher is established outside the EU,
the EU authorized representative is:
Springer Nature Customer Service Center GmbH
Europaplatz 3, 69115 Heidelberg, Germany

Printed by Libri Plureos GmbH
in Hamburg, Germany